P9-DNW-498

Dec 18

ASTRO-NAUT
AQUA-NAUT

Jennifer Swanson

NATIONAL GEOGRAPHIC

Washington, D.C.

ASTRONAUT

AQUANAUT

TABLE OF CONTENTS

Space has been called the "final frontier." It is certainly the most challenging frontier open to humankind, one that pushes us to explore the limits of our lives and learn how to sustain ourselves in the most demanding of environments. That's also why it is the best playground in the universe for human curiosity and imagination, the perfect place for us to exercise our natural instinct to explore and understand the world around us.

Here on Earth, we live in a tiny sweet spot in the universe. Our home planet has just the right mix of chemistry, air pressure, temperature, and sunlight to support a vibrant biosphere, including human life. Pushing human life beyond the surface of the Earth challenges us on every level. To survive in space, we must expand both scientific knowledge and our imagination and apply our new insights to develop innovative technologies.

When I was growing up, TV and magazines were full of stories about the people who were taking the very first steps into space, imagining and building the technology to live and work there right before my eyes. I was captivated by their personal stories, amazed by the complexity of things they had to figure out, and dazzled by their cleverness. Most of all, I was envious of the grand adventure they were having. I wanted to be a part of that grand adventure, to have a life as full of

creativity, innovation, and courage as the early space scientists and engineers. Achieving my childhood dream took a commitment to continual learning and a lot of hard work, but the life of discovery and adventure I've had far exceeds my youthful imaginings.

Space exploration pushes us to reach for our creative limits, and brings us a deeper understanding of ourselves and our place in the universe. The insights, lessons, and breakthroughs gained along the way also produce a rich cascade of advancement and benefits that improve life for many millions of people right here on Earth. Space exploration is, quite simply, one of the most spectacularly creative adventures available and one of the best drivers of progress for life on Earth.

—Kathryn Sullivan,
astronaut

For our species, exploration is arguably one of the biggest driving forces. It is what makes us who we are. This drive to push beyond the known world has taken on many forms. The necessity to find food, water, and shelter drove our prehistoric ancestors to expand their horizons. The quest by Europeans to find new trade routes to India allowed them to discover lands previously unknown. The drive by intrepid explorers such as Robert Peary, Frederick Cook, and Roald Amundsen to reach the poles of this planet in the last century pushed the boundaries of human endurance. Countless examples in every culture throughout the world exemplify our drive to reach the unknown.

Today we face a bit of a conundrum: Our species has come to a time when our basic needs are now met or, at the very least, are fairly easy to get. We have arguably become the dominant apex species—we have no predators anywhere on Earth—and have reached the farthest corners of the planet several times over. Technology connects each of us to nearly the entire human population; that's some 7.5 billion people on the planet. I often get the question "So what's left to discover?" from people of all ages. It's true—whatever you might want to explore, it seems like there is always someone who has done it before.

The good news is that there is so much left for those who want to push beyond the boundaries of the known! Part of the reason is that human beings have a tendency to think in a two-dimensional way when imagining adventure. In truth, we are at the dawn of modern-day exploration and have barely scratched the surface of both space and ocean discoveries.

My grandfather, Jacques-Yves Cousteau, was many things: a dreamer, a pioneer, an inventor, a filmmaker, an explorer, a conservationist, and much more. He opened up the undersea world to hundreds of millions of people. I, too, have an insatiable curiosity for the unknown. We have really only explored about 5 percent of our liquid realm on Earth. There are so many more mysteries; so many questions remain unanswered; so many strange new species are yet to be discovered! These are the kinds of things that keep an explorer's heart racing and mind yearning for more. This planet is our little blue oasis in space, our amazing life-support system, and we have the vast majority yet to discover lying in wait just beneath the ocean waves.

Astronauts and aquanauts share the same passion—to set off on a quest to learn more and to better ourselves.

—Fabien Cousteau,
aquanaut

INTRODUCTION:
UP AND DOWN

Deep space and the deep sea. You might not think they go together. They are two very different places. One is high above the Earth. The other stretches miles deep down below. One is a huge, empty vacuum, and the other is filled with water. Yet deep space and the deep sea are actually similar in many ways, too. The deep sea is a dark, mysterious environment with colossal mountains and bottomless pits. Places in outer space have those, too. The deep sea is home to volcanoes, earthquakes, and swirling vortexes. So is deep space.

Traveling in these two places is incredibly similar, too. Astronauts and aquanauts both have to worry about pressure, temperature, buoyancy, and most important, how to survive in a remote and somewhat hostile environment. **As dangerous as space and the deep sea sound, why would anyone go there? Exploration.** Humans are curious beings. We want to know more about the places around us, whether that means riding a rocket into space or submerging miles down into the ocean. Humans have already traveled to both places and brought back amazing information for scientists to study. But there is so much more to learn. There is a great need for new explorers. Scientists, engineers, mission control specialists, astronauts, and even journalists and teachers can all be part of an exploration team.

You, too, can become an explorer. Will you go up in space? Or down under the sea? Whether you choose space or sea, your mission is to learn all you can about each of these places. This book will give you great information on how to train for both environments. Check out the sidebars for tips on different things you might see in each place. Take a good look at the explorer's notebook for stuff you should do to prepare for your trip. Finally, read the words from actual astronauts and aquanauts about their experiences in deep space and the deep ocean. This information will help you understand how to survive in each climate.

Can't decide if you'd like to go to space or underwater? You can do both. Some of the planets in our solar system have seas so deep that satellites can't see the bottom. You could be a deep-sea diver and go into space! So if you want to climb a mountain on a new planet or slide down a rocky abyss into the deepest parts of the ocean, this is the place for you. **Join us on an amazing journey as we go up in space and dive deep down in the ocean** to explore the far-off places of our planet and the solar system.

EXPLORING NEW ENVIRONMENTS

gravity

pressure

buoyancy

pressure

gravity

Exploring deep space and the deep ocean isn't exactly a walk in the park. Neither place is warm and cozy. In fact, in both places, humans need to bring their own oxygen tanks just to survive. They also better wear a warm, protective suit. Space is cold! The vast depths of the ocean aren't any warmer. A few minutes unexposed in either place and a person could freeze to death. Pressure is also something every human explorer needs to understand. As you go deeper under the water, the pressure increases. But in space, there is no air, so there isn't any air pressure. That's good, right? Not exactly. Humans are used to having air push down on them on Earth. Without air pressure, our blood vessels would expand and most likely burst. That would not be a good thing.

THE **LIGHT SOURCE** FOR THE **OCEAN** IS ALSO THE **SUN.**

DARKNESS DESCENDS

Do you like the dark? Let's hope so. The first thing you'll notice in space or the deep sea is darkness. It is very dark in outer space. Unless astronauts are looking directly at the sun or see light reflected off an object, space seems completely dark. In order for our eyes to see light, it must bounce off of something and scatter. On Earth, the sun's rays bounce off the atmosphere. In space there is no atmosphere. That means that the light cannot be scattered. Instead, it is simply swallowed up by the vast emptiness.

>>>

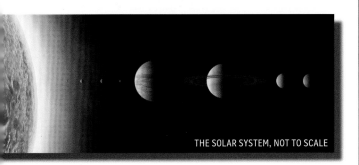

THE SOLAR SYSTEM, NOT TO SCALE

But wait, you *can* see things in space! That's because in space the light reflects off of the objects that are there: planets, asteroids, and other stars. If astronauts are looking directly at the sun, or at the Earth, which their spaceships typically face, they may see a lot of reflected light. That is because the Earth reflects the light from the sun. But in outer space, when there's nothing to reflect the light, it keeps traveling until it finds something. That is why we are able to see stars so far away. The "blackness" our eyes see is the absence of objects in space. That is why space appears black in photographs.

Just like space, the deep ocean is very dark. It makes sense. The light source for the ocean is also the sun. The sun, at its closest, is more than 91 million miles (146 million km) away from Earth. Despite the distance, the sun's rays penetrate the water, but only to approximately 656 feet (200 m). While that might seem like a long way, it's not. That's less than 2 percent of the ocean's depth. Tiny amounts of sunlight can sometimes reach as far as 3,280 feet (1,000 m). Light at this depth is very faint and appears as a dim greenish blue glow. But below that is complete darkness. The inky blackness of the ocean stretches for another 33,000 feet (10,000 m).

Like space, the deep sea does have occasional spots of light. This light comes from creatures or plants that give off their own glow. For the most part, though, space and the deep sea are both dark. So make sure to add a light source to your packing list. It might not do a lot of good in space, but at least in the deep sea, the light from your flashlight will find plenty of particles to bounce off of.

LIGHT FROM STARS SHINES OUT OF THIS GALAXY.

THIS DRAGONFISH GLOWS DEEP IN THE OCEAN.

WHAT GOES UP MUST
COME DOWN

Darkness is not the only thing you will notice on your journey into space. You will very quickly feel something unusual. You can float! It's weird because you can't float on Earth. Gravity is what keeps us grounded, quite literally. Gravity is a force of attraction between two objects. It acts on us every second by pulling us toward the center of the planet. That's a pretty good thing—without gravity, we'd float off into space! While you may not think of gravity much, you know it's there. Whenever you drop something and it falls to the ground, that is gravity at work.

How does it work? Every object in the universe has mass. Mass is the amount of matter an object has. It can be represented as pounds or kilograms. The mass of an object affects its gravity, or gravitational force. An object with a large mass has a larger gravitational force. What has

the biggest gravitational force in our space neighborhood? The sun. The sun is the most massive object in our solar system, and that means it has the greatest gravitational force. That force is responsible for keeping the planets in orbit. That's a pretty big job! Every planet has its own gravitational force, too. Earth's gravitational force is what keeps our moon in orbit around us, and the atmosphere in place over our planet.

But wait, you may have heard that space has zero gravity. That is not true. Space itself has gravity. It must. Gravity is the reason the planets came into existence. It was the pull of forces between bits of dust and matter that made them clump together. Gravity is the reason stars form clusters of swirling galaxies. Gravity is also the reason astronauts float. In space we call it microgravity.

NASA ASTRONAUT KAREN NYBERG FLOATS
INSIDE THE INTERNATIONAL SPACE STATION.

The Earth's **gravitational force** is what keeps our **moon in orbit** around us, and the **atmosphere in place** over our planet.

THE MOON OVER EARTH

"Micro" is another way of saying "very small," so microgravity refers to extremely small gravitational forces. In microgravity, objects are in free fall. That means they are constantly falling. Inside a spacecraft, astronauts, equipment, papers, and food packets are falling. Hold on. When you watch a video of an astronaut in a spacecraft, she doesn't look like she's falling; she appears to be floating. In fact, you've probably seen astronauts float balls or pens to each other. What you are seeing are not floating people and objects but falling people and objects, and this is caused by microgravity.

What is the MOST EXCITING part about being AN ASTRONAUT?

The most exciting part of being an astronaut is being able to fly everywhere and all the time. It is like being Peter Pan every day! I really loved being able to fly from place to place. In microgravity, we really can't walk … you have to fly!
—Cady Coleman, astronaut

The key to understanding microgravity is to know that space is a vacuum. No, not the kind you use to clean your house. In this case, the vacuum is a huge—mostly empty—area. Unlike Earth, there is no atmosphere or air in space. Without atmosphere, gravity cannot happen on a huge scale. Think of it this way: If you drop a hammer when you are standing on Earth, the hammer falls toward the ground. If you drop that same hammer while in space, it floats next to you. Why? You are falling at the same speed as the hammer. To your brain, it appears as if both of you are floating.

WHENEVER YOU **DROP** SOMETHING AND IT **FALLS TO THE GROUND,** THAT IS **GRAVITY** AT WORK.

But if everything falls in space, why doesn't the spacecraft fall out of the sky? As long as the spacecraft stays in space, and constantly accelerates at an appropriate speed, it will end up "falling" along a curving path around the Earth. This is called an orbit. The spaceship orbits the Earth high enough and just fast enough to avoid being pulled into the Earth's atmosphere. As long as the spacecraft keeps up its acceleration, it won't fall out of its orbit.

Microgravity is really helpful for astronauts, and not just for keeping them in space. Because of microgravity, moving a massive object in space is no problem! That's because with such little gravitational force, mass doesn't matter. Whether it's a huge satellite or a small hammer, it will float. All you have to do is direct it to the right place and give it a little push to send it where you want it to go. Makes you feel strong, doesn't it?

WHAT GOES DOWN MUST
COME UP

Dive underwater, and you will encounter another force in addition to gravity: buoyancy. It's pretty important to objects in the water. It's what makes them float on top of the water ... or sink. Just like gravity pulls objects toward the center of Earth—which makes them sink in water—buoyancy pushes up on an object to keep it afloat. The reason has to do with displacement.

BUOYANCY. IT'S PRETTY IMPORTANT TO OBJECTS IN THE WATER.

As a ship floats in water, it displaces, or pushes aside, an amount of water equal to its weight. The water that is left under the ship supports it, much like your arms would support a puppy or kitten you might be holding. So buoyancy has to do with weight. That means that objects with the same weight will all float. Right? Not exactly.

Buoyancy also has to do with density. Density is the amount of mass an object has in a given space. Objects that are made of a bunch of particles packed tightly together are dense. Think about a bowling ball. Objects where the particles are not packed tightly together are less dense. For example, a table tennis ball the same size as your bowling ball would be less dense than the bowling ball. It would float, while the bowling ball would sink.

Buoyancy also has to do with the surface area of an object. Take a huge rock and a ship of the same weight. The rock has a smaller surface area than the ship, since the bottom of the ship is wider and takes up more space than the rock does. This means that the weight of the ship is more spread out, and it has a greater surface

Just like **gravity pulls** objects **toward** the center of the **Earth**, **buoyancy pushes** up on an object to **keep it afloat**.

area. An object with a large surface area is more easily supported by the water. Thus, it floats. The idea works for an aircraft carrier with all its planes onboard. The huge surface area of the ship helps it stay afloat.

So both gravity and buoyancy affect an object floating, falling, or sinking. So how is gravity different from buoyancy? Buoyant forces act only in a fluid, such as water. Gravity forces can act anywhere: in air, in water, and even in empty space. In fact, as that ship floats in water, it feels gravity pulling down on it. The force of the gravity pulling down is balanced by the force of the buoyancy pushing up.

Just like understanding microgravity can help astronauts get the job done, the forces of buoyancy and gravity would be very important to you as a deep-sea explorer. You'd need to understand how your submersible will float and also how you will be able to get it to sink to extreme depths.

So how do you get a submersible, your very own underwater craft, to sink—and still come back up? You increase your craft's density to make it sink. The submersible's ballast system is the way to do that. The ballast system is equipped with special tanks. When the submersible needs to operate on the surface of the water, the tanks are filled with air. This makes the submersible lighter—and less dense—than the water around it, which means it floats. When the submersible needs to sink, tubes are opened in the tank and seawater streams in. As the water pours into the tanks, it makes the submersible heavier and increases its density. That makes it sink.

When the ballast tanks are filled with water, the submersible will be suspended underwater. The buoyant force acting on the craft is equal to the gravitational force acting on it. If the submersible wants to sink lower, it would need

SUBMERSIBLE

When a submersible needs to surface, the tanks are filled with air (1). This makes the submersible lighter and less dense so it can float. When the submersible needs to sink, tubes are opened in the tank and seawater streams in (2). The seawater makes the submersible heavier and increases its density. That makes it sink.

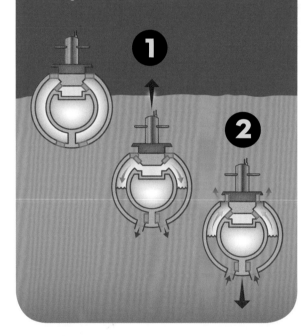

to use its engines to further push against the buoyant force.

Now how do you get back up to the surface? If you are relying on a ballast system, compressed air is forced into the tanks and the water is pushed out. As the water drains from the tanks, the submersible becomes lighter and less dense. That makes it slowly rise to the surface.

Need an easier system? No problem. Not all submersibles rely on ballast tanks. Some have weights attached to the outside of the craft that make them dense enough to sink.

UNDER PRESSURE

The last of the three forces you'll need to know about on your trip to space or sea is pressure. In fact, it's probably the most important of the three. Pressure is usually described as a force pushing on you. Astronauts need to be concerned about pressure because there isn't any in space. That is a very strange feeling. We humans are used to feeling pressure. Every day that we walk around on Earth we have 14.7 pounds per square inch (psi) (760 mm of mercury, or 1,013 millibars) pushing down on us. Don't worry. That pressure actually feels normal to you. But when you're outside of your spacecraft, that pressure will be gone. You will feel weightless.

SPACE SUITS NEED TO ADD ARTIFICIAL PRESSURE.

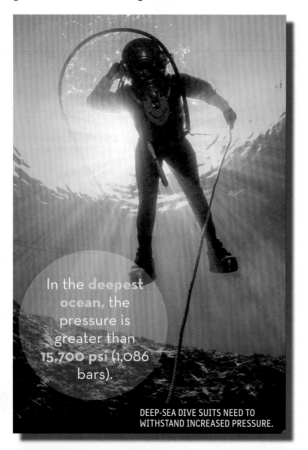

In the **deepest ocean,** the pressure is greater than 15,700 psi (1,086 bars).

DEEP-SEA DIVE SUITS NEED TO WITHSTAND INCREASED PRESSURE.

That's why all spacecraft and space suits are pressurized. The pressure within the spacecraft is set to normal atmospheric pressure. This allows the astronauts to live and work as normally as possible. The pressure in a space suit is set to .29 atm, less than that inside the craft, about one-third normal pressure. This is necessary because when your body is used to having pressure, not having it is not good. Without pressure pressing down on your body from all sides, your insides would expand outward (your internal organs would move apart) because there is no force holding them together.

While space has no pressure, the deep sea has tons of it. As you go down in the ocean, pressure increases. For every 33 feet (10 m) you descend, the pressure on your body doubles. In the deepest ocean, the pressure is greater than 15,700 psi (1,086 bars). That is a thousand times the amount of pressure you feel when standing on the ground. To give you a better idea of what it would feel like, imagine one person holding up 50 jumbo airplanes. That is a lot of pressure!

Pressure under the water is related to buoyancy. As the pressure increases, so does the buoyant force. In order for a submersible to stay submerged as it goes deeper, it must have more and more weight pressing down on it to keep it moving downward. Otherwise, the buoyant force would be greater and would work to push the submersible back up.

Pressure doesn't just push in on you. It also affects how your body works. As pressure increases, your body absorbs gases differently. Normally, you have oxygen floating around in your cells. Oxygen is what we need to breathe and survive. When you dive deep, your body absorbs more oxygen. It also absorbs other gases, too, like nitrogen. These gases collect in your tissues and are a normal result of diving. But before coming back to the surface, you need to get rid of that excess nitrogen.

If you don't, you could get a condition called the bends. The bends can make a diver feel pain and tingling or even loss of feeling in an isolated part of the body. And in extreme cases, the bends can cause death. To stay healthy while diving, take your time as you go back up. This allows the excess nitrogen to slowly dissolve out of your tissues and to be eliminated from your body. This process can take several hours or even several days.

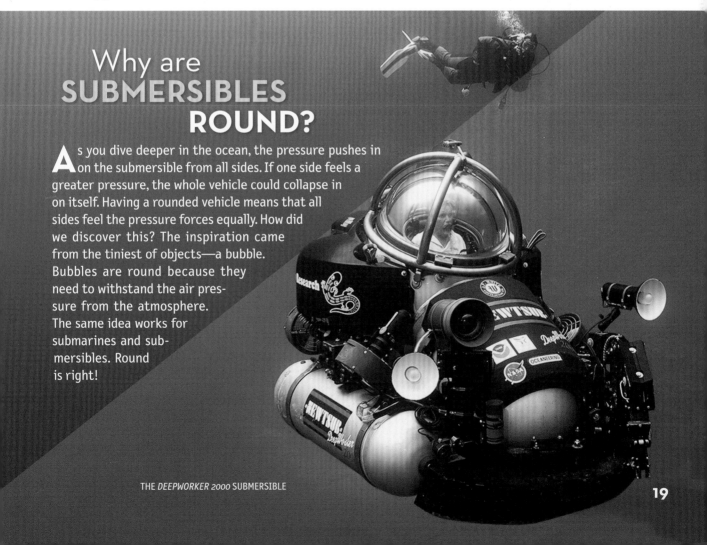

Why are
SUBMERSIBLES
ROUND?

As you dive deeper in the ocean, the pressure pushes in on the submersible from all sides. If one side feels a greater pressure, the whole vehicle could collapse in on itself. Having a rounded vehicle means that all sides feel the pressure forces equally. How did we discover this? The inspiration came from the tiniest of objects—a bubble. Bubbles are round because they need to withstand the air pressure from the atmosphere. The same idea works for submarines and submersibles. Round is right!

THE *DEEPWORKER 2000* SUBMERSIBLE

TEMPERATURE, TOPOGRAPHY, AND PLATE TECTONICS

If you're going to go up in space or dive deep in the ocean, you will need some new clothes to wear. Specifically, you'll need a special suit. Why? First of all, it will be equipped with air so you can breathe. It will be adapted to produce a pressure that you can survive in. It will also be insulated because ... it's cold! Both space and the deep sea are frigid. The temperature deep in the ocean can be as cold as 28°F (-2°C). That is below freezing. Considering that the normal temperature for a swimming pool is between 83 and 86°F (28 and 30°C), that's pretty cold!

Space is much colder. In space, when the sun is not shining on you, it can drop to minus 148°F (-100°C). You'd freeze instantly if you were exposed to that temperature. That's why space suits have heaters. They have tiny coolers, too. That's because in the sunlight, without an atmosphere to filter the rays, it gets hot. Too hot for humans. How's 500°F (260°C) sound? That's just a bit toasty. Bet that space suit is looking good right about now, huh? Suit up!

Once you have your suit on, you are ready to go out and explore. Whether you are going on your first space walk or diving deep under the water, you need to take time to look around. You may just see some familiar topography. Topography is how the land looks. There are huge mountains deep underwater, and there are huge mountains on some of the planets in space.

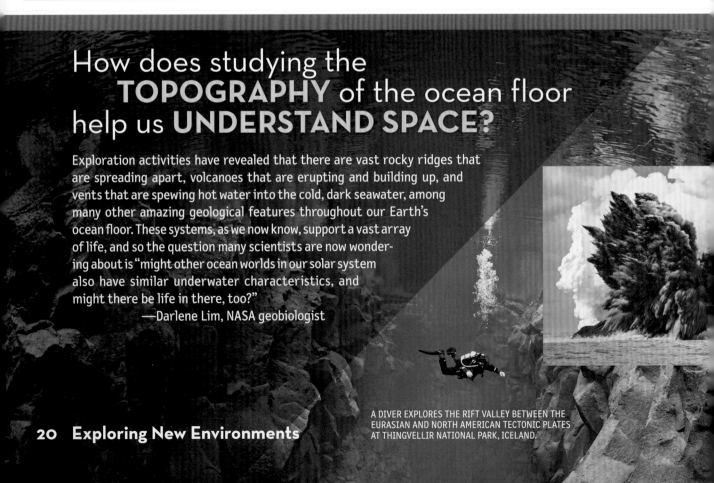

How does studying the TOPOGRAPHY of the ocean floor help us UNDERSTAND SPACE?

Exploration activities have revealed that there are vast rocky ridges that are spreading apart, volcanoes that are erupting and building up, and vents that are spewing hot water into the cold, dark seawater, among many other amazing geological features throughout our Earth's ocean floor. These systems, as we now know, support a vast array of life, and so the question many scientists are now wondering about is "might other ocean worlds in our solar system also have similar underwater characteristics, and might there be life in there, too?"
—Darlene Lim, NASA geobiologist

A DIVER EXPLORES THE RIFT VALLEY BETWEEN THE EURASIAN AND NORTH AMERICAN TECTONIC PLATES AT THINGVELLIR NATIONAL PARK, ICELAND.

OLYMPUS MONS, MARS

Mars has some of the **highest mountains** in the solar system.

MOUNT EVEREST, EARTH

Mars has some of the highest mountains in the solar system. The highest, Olympus Mons, tops out at over 15.5 miles (25 km). That is almost three times as high as Mount Everest! The tallest underwater mountain is Mauna Kea, an inactive volcano that is part of Hawaii. The mountain is so high that more than 13,700 feet (4,200 m) of it appears above the surface of the ocean, but almost one and a half times more land stretches into the depths of the Pacific Ocean. The total height is more than a mile (1.6 km) higher than Mount Everest.

SOME VOLCANOES EXIST FULLY UNDERWATER. THIS VOLCANO'S MASSIVE ERUPTION CAN BE SEEN ABOVE THE SURFACE, OFF THE COAST OF TONGA.

WHIRLPOOL

Look at this whirlpool. It is a swirling mass of water that pulls things into it, like people and boats. Now imagine this in space. There, it is called a black hole. Black holes are massive objects that pull everything toward them: light, spaceships, asteroids. Scientists don't know exactly what causes whirlpools and black holes, or whether they are even related, but they imagine them to be similar. What does that mean to you? Steer clear of anything swirling!

AN ARTIST'S INTERPRETATION OF A BLACK HOLE

NASA's Mars **Reconnaissance Orbiter** took this photo of an **impact crater** on Mars using the Orbiter's High Resolution Imaging Science Experiment (HiRISE) camera.

>>>

Craters, too, are found in both the ocean and on planets. The Burkle Crater in the Indian Ocean spans more than 18 miles (30 km) across. In space, the largest crater found so far is on Mars and is more than 1,400 miles (2,300 km) in diameter.

You can also check out moving land both in the ocean and in space. Earthquakes are the result of the shifting of huge sections of land called tectonic plates, which make up Earth's crust. These

plates are irregularly shaped slabs of rock that fit together like a giant puzzle. Sometimes the plates move slightly or shift and scrape against each other. This causes the landforms above to move, too, resulting in an earthquake. During an earthquake, the ground shakes, buildings can topple, and landslides occur. Under the ocean, an earthquake may create a giant wave called a tsunami. Tsunamis can have waves that reach up to 30 feet (9 m) and are usually quite devastating

to coastal cities. Sometimes movement of tectonic plates will form volcanoes. When the plates move apart or push together, the extremely hot rock, or magma, shoots up through the cracks and comes out as lava. Volcanoes and earthquakes don't happen just here on Earth. Scientists have observed earthquakes and volcanoes in many spots in space, including on Mars, Mercury, and Io, one of Jupiter's moons.

EXPLORER'S NOTEBOOK

1 LEARN ABOUT THE ENVIRONMENT

✓ Gravity and Microgravity

✓ Buoyancy and Density

✓ Pressure and Temperature

✓ Topography

ACTIVITY:

SINK OR FLOAT

Submersibles need to be able to sink and float at a certain depth.

HOW CAN THEY DO THAT?

It has to do with pressure.

SUPPLIES

A pen cap, otherwise known as your "submersible"

A piece of clay or putty that is waterproof

Large 2-liter bottle filled with water

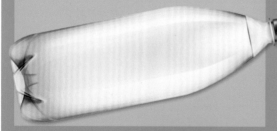

Cap to the 2-liter bottle

STEPS

1 Take the pen cap and place it in the filled 2-liter bottle. Notice how it floats at the top of the water.

2 Remove the pen cap. Add a small amount of clay or putty to the long end of the pen cap.

3 Place the pen cap into the bottle. It should float much lower in the water, but it should still float.

4 Now put the cap on the bottle.

5 Squeeze the bottle on both sides.

6 Watch your "submersible" sink. As you apply pressure by squeezing, you are creating forces similar to the ones in the ocean. The pressure from the bottle makes the air caught in the top of the pen cap shrink. That increases the density of the pen cap and it sinks.

7 Stop squeezing the bottle.

8 Your submersible pops back up to the top. This is because the pressure is less, and it takes more weight to sink at a lower pressure.

THINGS TO THINK ABOUT:

What would happen if you **added too much weight** to the pen cap?

How would this work if it were **a diving bell** that fit over a human head?

What kind of **pressure forces would a diver feel** as she descends below 33 feet (10 m)?

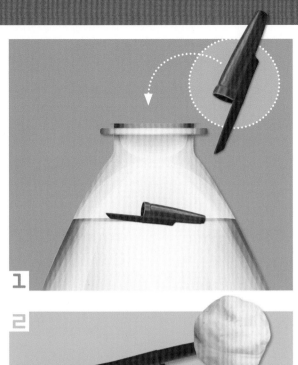

1

2
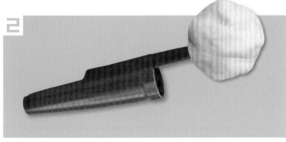

3
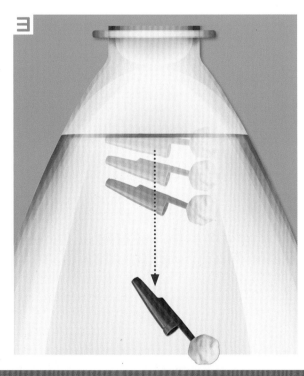

4

5,6
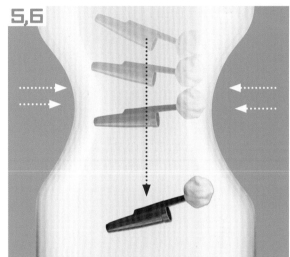

7,8
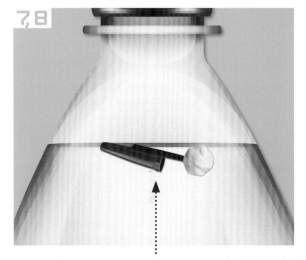

CHAPTER 2

BLASTING OFF

OR TAKING THE PLUNGE

epipelagic zone
mesopelagic zone

bathypelagic zone

abyssopelagic zone

hadalpelagic zone —

exosphere

thermosphere

mesosphere

stratosphere

troposphere

Still interested in becoming a space or deep-sea explorer? Good. Congratulations! You have been accepted into the astronaut/aquanaut training program. For the next two years, you will learn the basics of flying, diving, and being part of a team environment. All of these are extremely important for your success. Let's get started.

ASTRONAUT-IN-TRAINING

As an astronaut-in-training, you will spend many hours preparing for your journey. Of course, you'll learn about the science of going into space. You'll find out how pressure and temperature will affect you. Plus, you will receive medical training so that you understand how your body will change in microgravity. For example, you actually grow. Since gravity is not pushing down on you, the space between your vertebrae—the discs in your back—stretches out. You can get up to three inches (7.6 cm) taller! Your hair may grow more quickly. Your heart will beat at a slower rate, and your muscles will get weaker. Muscles here on Earth are needed to continually overcome gravity. You use muscles to walk up big hills, climb stairs, and run around the block. In space, there is only microgravity, so an astronaut's muscles get weak from not being used.

During your preparation you will receive not just normal medical training, like how to take care of cuts and burns, but also emergency training. If someone has an immediate medical problem in space, there isn't going to be a rescue squad coming up anytime soon.

Astronauts also go through survival training. That means they spend three days in the wilderness with very few supplies. They must make their own fires, find and cook their own food, and create a safe place to sleep. Sounds harsh? Maybe. But astronauts are subjected to these trials so that they learn to cope with any type of situation. What if their spaceship is forced to land back on Earth on a tropical island, a high mountain, a desert, or even a glacier? Will they survive? With their training in climbing, descending high cliffs safely, crossing rivers, and using the stars for navigation, their chances are much better. Astronauts are also expected to be good swimmers and to become certified scuba divers.

NASA'S ASTRONAUT REQUIREMENTS

 Graduate from college with a degree in science, technology, engineering, or math (STEM)

 Three or more years of experience in a STEM field

 Ability to pass the astronaut physical

Wait a minute! They're going into space, not into the water. Why do they have to know how to swim? Well, one reason is that their spacecraft may have to make an emergency landing in the water. The other reason has to do with gravity.

ASTRONAUTS BEING INTRODUCED TO CLIMBING TECHNIQUES BY A PROFESSIONAL MOUNTAINEER DURING SURVIVAL TRAINING

It's extremely difficult to train for the microgravity environment in space here on Earth. After all, you can't get away from gravity. For years, the only way that astronauts-in-training could experience "real" microgravity was through a pretty crazy plane ride. NASA has a KC-135A turbojet called the "Weightless Wonder" that allows astronauts to float for 25 seconds. How does it work? The plane flies up to 26,000 feet (7.9 km) and then starts an extremely steep climb curving upward in an arc shape. It races upward as fast as it can. The people inside feel a huge pull of gravity, more than 1.8 times the normal amount. As the plane curves downward and goes into sort of a free fall, astronauts free-fall as well. They float for up to 25 seconds and then the plane levels off. It's a short time to experience weightlessness, but it gives the astronauts a tiny idea of what it will be like in space.

Discovering Operation NEEMO

Astronauts go up in space. They also go deep under the water. While it may seem strange, NASA trains its astronauts in a facility that has long been used by aquanauts. The NASA Extreme Environment >>>

It's very expensive to use the turbojet, and the 25 seconds of microgravity is not long enough to really understand how it will affect you. So, NASA turned to the only other place on Earth where you can get away from gravity: the ocean. Gravity is still a factor underwater, but buoyancy offsets gravity. That makes underwater the perfect place for astronauts to train. It's the only place on Earth where the microgravity environment in space can be easily duplicated for long periods of time. Training underwater makes sense for other reasons. Both places are dark, isolated, cold, and most important, without breathable oxygen.

The last part of astronauts' training involves learning about the environment where they will be living and the equipment they will be using.

Astronauts study maps of the International Space Station (ISS) to know each compartment and how it is used. They learn how to communicate with Earth, how to make their own food, how to set up their bed, and even how to go to the bathroom in space. Astronauts may also be asked to learn a different language so that they can communicate with their international partners on the ISS.

Why is it IMPORTANT for astronauts to TRAIN UNDERWATER?

When you are living underwater during NEEMO missions, you are far away from your family for weeks at a time, living with only a few other people, day and night, in a place that is about the size of a school bus. This means that you are learning not only how to work together but also how to live together as a team and how to look after each other when you are tired or hungry or upset about something. And at all times, especially anytime you leave your underwater home to explore the ocean floor, you have to make sure that all of your equipment is working perfectly, that you always follow safety rules, and that you always know what to do in an emergency, because any mistakes can be very, very dangerous. Just like being on a space mission.

—Dr. Andrew Abercromby, NASA engineer

> > > Mission Operations team, known as NEEMO, spends three weeks training in Aquarius, an underwater research center submerged off the coast of Florida. Astronauts work with trained aquanauts and have experiences onboard Aquarius similar to the ones they will have on the International Space Station. Both places are remote and have very tight living quarters and unique living arrangements. The aquanauts are able to mimic many tasks they will be asked to perform in space. They don suits and go out of the research vessel to "spacewalk" on the floor of the ocean. They maneuver large objects into place and learn to rely on teamwork to get things done. What a great way to get your feet wet!

DEEP-SEA DIVER OR
AQUANAUT-IN-TRAINING

Unlike for astronauts, there isn't really a specific training path for a deep-sea diver. Deep-sea divers who work as scientists will spend many years in school learning about their field of study. They can be marine biologists, oceanographers, geophysicists, geobiologists, marine conservationists, and many other types of scientists.

Every deep-sea scientist should be a certified scuba diver. That means that you will need to take classes that teach you how to operate your scuba equipment. You'll learn how pressure acts on the body as you descend in the water. You should be able to confidently dive in all different types of environments: deep water, open water, and at different depth levels. Scuba divers should be in good physical shape and be great swimmers. In order to live in Aquarius, the only underwater laboratory, aquanauts must have successfully completed at least 50 scuba dives.

Working underwater for long periods of time will be similar to working in space. You'll be in a small area with a few other people, or possibly

PREPPING for Your DIVE

1

CHECK YOUR TANK.
YOU NEED TO MAKE SURE THAT THE TANKS ARE FULL OF OXYGEN SO YOU'LL HAVE ENOUGH AIR TO BREATHE.

2

GEAR UP.
PUT ON YOUR TANK, MASK, AND FINS. THEN GIVE YOUR REGULATOR A TRY. YOU WANT TO BE SURE THAT IT IS WORKING SO YOU WILL GET AIR.

3

TALK WITH YOUR PARTNER.
SET UP SIGNALS FOR HOW YOU WILL COMMUNICATE. ALSO DETERMINE HOW LONG YOU WILL STAY DOWN AND WHAT TO DO IF YOU HAVE AN EMERGENCY.

NOTE: YOU'LL NEED FORMAL SCUBA TRAINING FROM AN OFFICIAL INSTRUCTOR!

YOU SHOULD BE ABLE TO CONFIDENTLY DIVE IN ALL DIFFERENT TYPES OF ENVIRONMENTS: DEEP WATER, OPEN WATER, AND AT **DIFFERENT DEPTH LEVELS.**

AQUARIUS AQUANAUT REQUIREMENTS

✓ Graduate from college with a degree in STEM

✓ Must have already worked underwater for 24 hours

✓ Be scuba qualified and have successfully completed 50 dives

✓ Ability to pass the aquanaut physical

alone. You'll be solely dependent upon your equipment for food, water, oxygen, and a pressurized living space. And you'll become accustomed to complete darkness or artificial lights from within the cabin.

4 TAKE THE PLUNGE!
STEP INTO THE WATER AND START SWIMMING. ENJOY WATCHING THE AQUATIC WORLD ALL AROUND YOU.

CUTTING THROUGH
THE LAYERS

You now understand that going into space or deep-sea diving are similar. They're both dark and have very small spaces, and not many people can go there. It is hard work to be an astronaut or aquanaut. It's best if you don't get seasick—or airsick. You can avoid that by learning to adapt to changing pressures and temperatures, and you need to be in top physical shape. Finally, both jobs require a commitment of time. Time for training. Time for practicing. Time away from your family. But what will you experience when you actually go up in space or down in the ocean?

GOING UP

Whether you are going up or down, your craft will have to pass through many different layers. For the most part, you may not even notice the layers. After all, they are invisible. But each layer of the atmosphere or the ocean is separate and has its own properties.

The atmosphere above the Earth is made up of five layers. The thickest layer starts at ground level and stretches up about 33,000 feet (10 km) high. It is called the troposphere, and it's the one that we live in. The height of the troposphere can change depending on where it is over the Earth and what time of year it is. During the summer, the troposphere extends higher. In the winter, it's lower. Near the poles, the troposphere tends to be lower, while it's higher near the Equator. The troposphere contains 99 percent of the water vapor in the atmosphere. That makes it home to the weather system that we feel here on Earth. Rain, snow, sleet, wind, and even tornadoes and hurricanes all happen in the troposphere. The air in the troposphere is warmer near the surface of the Earth and gets colder as you go up. Air pressure and density also decrease as you go higher. That's why the cabins in an airplane are pressurized, so you can still breathe normally.

The next layer up is the stratosphere. It extends 31 miles (50 km) above the surface of the Earth. The stratosphere contains one of the most important safety measures of our planet: the ozone layer. This layer is a thin shell of ozone, a special coating of gases that reflect some of the incoming rays from the sun. It's kind of like a blanket that keeps the Earth warm by trapping some of the heat from the sun close to the planet's surface. But it also acts like sunscreen, reflecting some rays back out in space. We don't want to be too warm, right? Unlike the troposphere, the stratosphere is very dry. There aren't many clouds in this part of the atmosphere.

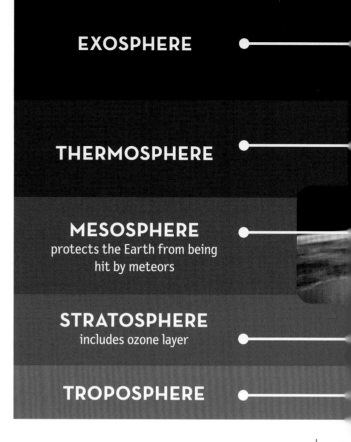

EXOSPHERE

THERMOSPHERE

MESOSPHERE
protects the Earth from being hit by meteors

STRATOSPHERE
includes ozone layer

TROPOSPHERE

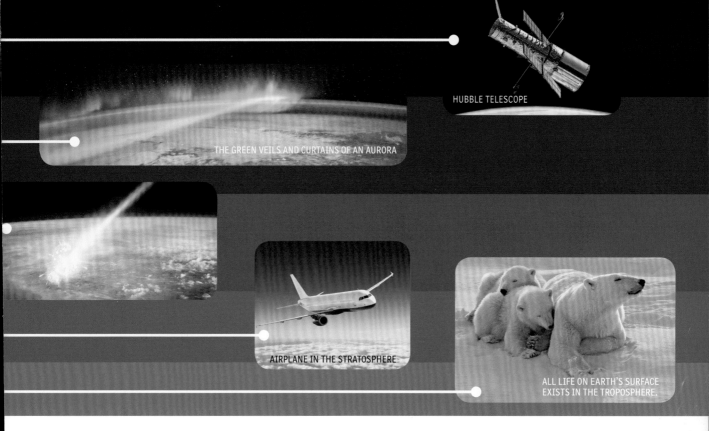

HUBBLE TELESCOPE

THE GREEN VEILS AND CURTAINS OF AN AURORA

AIRPLANE IN THE STRATOSPHERE

ALL LIFE ON EARTH'S SURFACE EXISTS IN THE TROPOSPHERE.

Because of the lack of weather, the stratosphere is the perfect place for planes to fly. Without all the precipitation and wind, the ride is less bumpy. Air in the stratosphere is one thousand times thinner than at the surface of the Earth. That makes breathing without an oxygen tank impossible. Pressure is very low here, too. If humans were to travel outside of a pressurized craft into the stratosphere, they'd need pressurized suits.

The third layer of the atmosphere is the mesosphere. It spans an area from 31 miles (50 km) to 53 miles (85 km) above the Earth. The mesosphere is the coldest layer. The temperatures here are less than minus 130°F (-90°C)! This is the layer where most meteors burn up before they can get close to the Earth.

The fourth layer up is where you will find the thermosphere. It extends from about 56 miles (90 km) to between 311 and 621 miles (500 and 1,000 km) above Earth. The thermosphere is

really hot! That's because it feels the direct heat from the sun's rays. Temperatures can be as high as 3632°F (2000°C) or higher. Since "space" begins at about 61 miles (100 km) above Earth's surface, this is the most important layer for you. Your spaceship will go into orbit in the middle to upper layers of this area. You will be in a perfect spot to view the northern lights (aurora borealis) in the night sky. But be sure to watch out for satellites and space garbage. They hang out in this layer, too.

The highest layer of the atmosphere is the exosphere. It is where true outer space starts. The atmosphere is extremely thin here, and the temperatures are excessively high. It can be as hot as 4530°F (2500°C) on a sunny day. At night, however, it can be very, very cold. You will find a few high-orbit satellites here as well as the Hubble Space Telescope.

GOING DOWN

Just like the atmosphere, the area under the ocean has five layers, too. Underwater layers are called zones. The surface layer is called the epipelagic zone, but it's better known as the sunlight zone. That's because the rays of the sun are able to penetrate all of its 656 feet (200 m). That is a good thing for the plants and animals that are found here. The heat from the sun also keeps this zone fairly warm, although pockets of cold water do still exist here. Snorkelers, scuba

divers, surfers, and sailboats are all found here. The sunlight zone is where the ocean connects with the land, too.

The next layer down is the mesopelagic zone. It extends from the bottom of the epipelagic zone to between 2,300 and 3,280 feet (700 and 1,000 m) deep. This zone is sometimes called the twilight zone because very little light penetrates this deep. Twilight is the time of day when the sun is almost down. There are still faint rays of

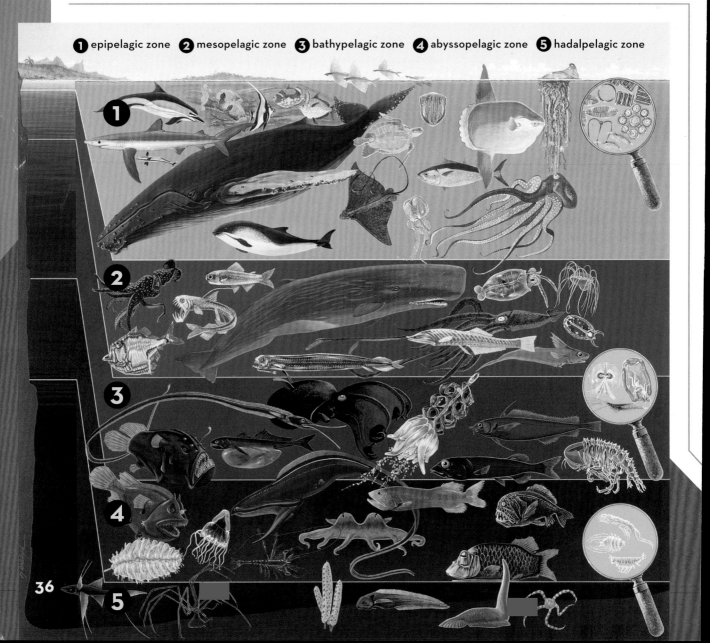

1 epipelagic zone **2** mesopelagic zone **3** bathypelagic zone **4** abyssopelagic zone **5** hadalpelagic zone

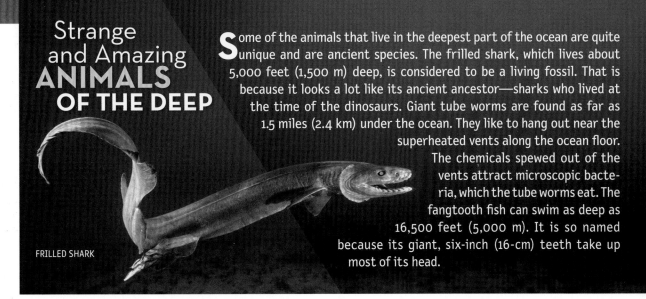

Strange and Amazing ANIMALS OF THE DEEP

Some of the animals that live in the deepest part of the ocean are quite unique and are ancient species. The frilled shark, which lives about 5,000 feet (1,500 m) deep, is considered to be a living fossil. That is because it looks a lot like its ancient ancestor—sharks who lived at the time of the dinosaurs. Giant tube worms are found as far as 1.5 miles (2.4 km) under the ocean. They like to hang out near the superheated vents along the ocean floor. The chemicals spewed out of the vents attract microscopic bacteria, which the tube worms eat. The fangtooth fish can swim as deep as 16,500 feet (5,000 m). It is so named because its giant, six-inch (16-cm) teeth take up most of its head.

FRILLED SHARK

the sun stretching across the sky. That's kind of what this zone is like. Submarines silently glide through the cold waters of the twilight zone. Periodically, glowing creatures light up the water, providing light where there is none. The glow comes from bioluminescence, the ability of an animal to produce its own light inside its body.

The next three zones are completely dark. The only visible light in these areas comes from bioluminescent creatures. The bathypelagic zone is the third area down from the surface, extending from 3,300 to 13,100 feet (1,000 m–4,000 m). Here, in the dark depths of the ocean, pressure is intense. It's so strong that your lungs would instantly collapse. You couldn't breathe. How do animals live there? Some have lungs that can collapse if necessary. Sperm whales' lungs collapse to keep them from bursting under the pressure. That allows the whales to be able to hunt for squid at more than 7,000 feet (2,100 m) under the sea. At 41°F (5°C), the bathypelagic zone is cold, too.

It's not the coldest part of the ocean, however. That title belongs to the abyssopelagic zone, or the abyss. At depths from 13,100 to 19,700 feet (4,000–6,000 m) deep, the temperatures can be barely above freezing. The abyss is the largest environment on Earth, covering more than 115 million square miles (300 million sq m), or about 60 percent of the Earth's surface. The abyss accounts for more than 75 percent of the ocean and is one of the most difficult places to explore on the planet. Why? With its extreme cold and the gigantic pressure it puts on a human, it's very difficult to dive that deep. The pressure a human would feel in the abyss is equal to having two elephants on top of each other stand on a postage stamp. Talk about a high-pressure environment!

THE **ABYSS** IS THE **LARGEST ENVIRONMENT** ON THE **EARTH.**

The last zone in the deep sea is the hadalpelagic zone, or the hadal zone for short. This zone reaches areas that are more than 19,680 feet (6,000 m) deep, and are found only in large trenches on the ocean floor. The Mariana Trench stretches 36,160 feet (11,021 m) below the surface. Very few people have made it to the Mariana. In fact, more people have been to the moon (twelve) than have been to the Mariana Trench (one).

COUNTDOWN TO
BLASTOFF

It's time. You have completed your training and now you are ready for your mission into space. You and two of your crew members climb into the Soyuz space capsule. It's a tight squeeze, but each of you has your own seat. You fasten your harness, making sure it's good and tight. As you sit for two and a half hours waiting for blastoff, you review all of your training. Are you ready? Yes! Finally, the capsule begins to rumble. The engines have started. 3 ... 2 ... 1 ... Liftoff!

The rocket propulsion system ignites and you stream upward, leaving behind huge billows of white steam that look like fog. You are pushed deep into your seat by the massive gravity forces. The capsule continues to shake and vibrate. You feel pinned to your seat, unable to move. With a big bang, one engine falls away and you are thrown forward, then immediately pinned back as the rocket pushes upward. The gravity forces are now close to four times normal. You feel like an elephant is sitting on your chest. The pressure grows for another nine minutes. Then, all of a sudden, it's gone. You feel weightless. You have just reached the edge of space.

The weight and fastening of the harness are the only things keeping you in your seat. Your arms might just float up all on their own. Welcome to space and microgravity.

But wait, where's the space station? Shouldn't it be sitting there waiting for you? Not exactly. Your rocket took only 10 minutes or so to reach space, but sometimes it can take hours to meet up with the space station. One reason is that your spaceship is at a lower level of orbit than the ISS, which means that the ISS is higher than you. Imagine that you're on the

ground and your friend is up in a tree. You're too far to stretch out your hands and connect. The same thing happens with your spacecraft and the ISS. You must steer your ship into the higher orbit to meet up with the ISS. It's not as easy as it sounds. You can't just make a right turn and get there. You're in a circular orbit. That means you need to nudge your way up slowly. Think of it this way. A running track has different lanes. You are in the closest lane and need to get to the outer lane. The only way to

THREE ... TWO ... ONE ... LIFTOFF!

What does it FEEL LIKE during BLASTOFF?

Blastoff is very exciting. There is a rumble and roar associated with rocket ignition, plus I knew that eight and a half to nine minutes later, I would be in the weightlessness of space, looking down at the Earth.
—Norman Thagard, astronaut

shift one lane is to go around the track one time. Each time you move a lane, you get a boost from your engines. If there are six lanes, it could take you six turns around the track to get to the outer lane.

Once you get there, the other problem is that the ISS is moving at more than 17,000 miles an hour (27,350 km/h). Your spaceship needs to catch up to the ISS so that it can dock, or attach, to it. The best way to do that is to let the ISS catch up to you. As you move into the same orbit, make sure your ship is in front of the ISS so it can catch your spaceship from behind. Whew! Was that tiring or what? Sometimes it can take as long as six hours or more to dock with the ISS. The whole time you are sitting in a cramped spaceship.

DIVE! DIVE!
DIVE!

You climb into your submersible. It's a pretty tight fit—just you and maybe one or two others. You wriggle around in your seat to get comfortable. After all, you'll be sitting there for the next eight to twelve hours or so. The giant crane on the research vessel lifts your submersible in the air and slowly lowers it into the ocean. Splash! You watch as the water level on the window in front of you rises. Until ... you are submerged. The thick cables attached to the top of the craft continue lowering you slowly through the water. At the rate of 115 to 131 feet (35–40 m) per minute, it can take up to two hours to reach your maximum depth of almost 20,000 feet (6,000 m). It's okay, you're not in a rush. Although you're descending into depths where the pressure outside of your submersible would crush you, you really don't feel it. The inside of the submersible is kept at a constant atmospheric pressure, which is the normal pressure

WHAT TO PACK:
ASTRONAUT

- Running shoes for the treadmill and bicycle
- Exercise clothes (T-shirt and shorts)
- White T-shirts to wear under work shirts (it's a standard part of the uniform)
- Shirts, shorts, and pants for work
- Underwear and socks
- Sweaters
- Space suit
- Personal items like pictures, stuffed animals, or special mementos

DEEP-SEA SCIENTIST

- Clothes similar to the astronaut
- Additional tools, such as zip ties, screwdrivers, pliers, pocketknife, duct tape
- A few personal items, such as an MP3 player for listening to music, pictures and special mementos
- Deep-sea scuba gear

Remember, there isn't a washing machine in either place. There's no way to wash clothes. Be sure to take clothes you like. You'll be wearing them a long time!

MIR DEEP-SEA SUBMERSIBLE

here on Earth. At most you may feel like you do when you are taking off in an aircraft: a slight pressure in your ears and tightness in your chest. As you settle deeper under the water, you become used to it, and that feeling goes away.

As you descend through the depths, the darkness consumes you. You turn on the lights outside your submersible. The six 5,000-watt lightbulbs would be blinding on the surface, but deep underwater the light hits the water and reflects back. You are able to see only a few feet in front of your craft. The water doesn't allow much light through its thick layers. Flash! A greenish blue creature swims by. Must be a squid or a jellyfish. It is so dark now that the only light

What does it FEEL LIKE while you are DESCENDING or ASCENDING in the OCEAN?

It is quite cold in the deep. When someone dives in a manned submersible, the chamber starts out warm and becomes much colder after hours in the deep. This causes water to condense and drip on the people in the chamber.

—Michael Vecchione, aquanaut

When you ascend, it is a slow glide up to the surface. You are anticipating getting to the top since you have probably been in a cramped position for up to eight to twelve hours in your submersible.

—Kathryn Sullivan, astronaut/aquanaut

comes from the lights on your submersible or from creatures that glow with bioluminescence.

Once you reach your maximum depth, the pilot turns on the hydraulic thrusters and you plunge smoothly forward. Your speed is still slow at only one to five knots, but that is good. You want to get a good look out of your viewing window. Side propellers allow the pilot to steer the craft back and forth. Having an adjustable ballast tank means that the submersible can go up and down. The ease of movement allows the pilot to hover over interesting features just like a helicopter, but without the wind blast from the blades. Your submersible is equipped with a robotic arm that you can maneuver via controls. It's great for capturing specimens for your research. The video camera takes excellent movies of everything you see. Still, you take notes furiously in your notebook, writing down everything that swims by. After all, you have only two hours down here.

Typical dives with a submersible last no longer than ten hours, but to be safe they are usually kept to about six hours total. Since you only have two hours at the deepest depth, you need to get all the information you can as quickly as possible. Coming back is not easy. In a short 120 minutes, it's time to go. As you sit back and relax for the two-hour climb back to the surface, you reflect on your amazing trip to the most unexplored portion of the planet.

HEADING DOWN FOR A LONGER STAY

If your plan is to stay longer underwater, you will be heading to the only underwater research facility in existence. Aquarius is a habitat that was created for aquanauts to live in for at least one day or even several weeks underwater. It is reached by scuba diving through the water and entering the "wet porch." So pull on your goggles, put on your flippers, and suit up! Take a backward dive into the ocean as you set off for your underwater home for the day, week, or month.

EXPLORER'S NOTEBOOK

2 TRAIN FOR THE TRIP

- ✓ Study
- ✓ Exercise
- ✓ Learn how your body will respond to space and underwater
- ✓ Decide what to pack

ACTIVITY:

DOCKING THE ISS

While it may look easy, having a spacecraft dock with the International Space Station is anything but. Imagine trying to place a baseball into a hole just barely big enough to hold it—all while you are floating around in space. That's right. You don't have anything to hold on to. **THINK YOU CAN DO IT?** Why not give it a try?

SUPPLIES

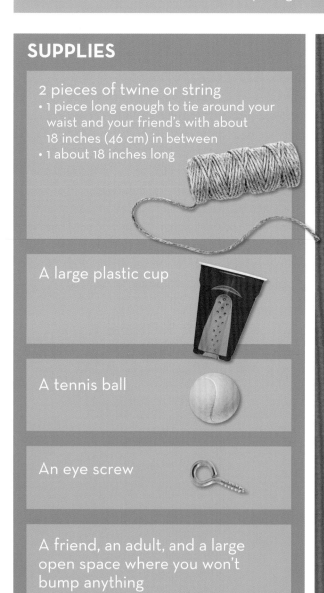

2 pieces of twine or string
- 1 piece long enough to tie around your waist and your friend's with about 18 inches (46 cm) in between
- 1 about 18 inches long

A large plastic cup

A tennis ball

An eye screw

A friend, an adult, and a large open space where you won't bump anything

STEPS

1 Have an adult help you screw the screw into the tennis ball.

2 Take the long twine/string and tie one end around your waist. Have your friend tie the other end around his or her waist.

3 Tie the smaller piece of rope to the long rope, about midway between you and your friend.

At the end of the small piece of rope, attach your tennis ball.

4 Set the cup on the ground between the two of you.

5 Step away from your friend until the long rope is taut, or straight across. The tennis ball should be hanging straight down.

Now, without using your hands, walk back and forth or move your body to maneuver the tennis ball into the cup. It may take a while. You will see that this requires a fair amount of cooperation between you and your friend.

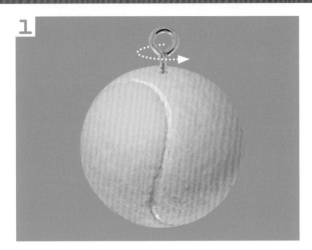

1

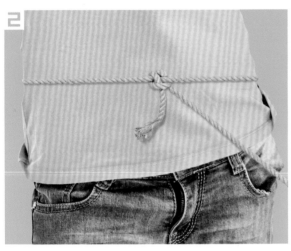

2

3

4, 5

THINGS TO THINK ABOUT:

What is the **relationship between your movement** and your friend's?

What happens when you and your friend **don't communicate** about your movements?

What other **forces or effects** would you need to consider in space?

Living areas
in a sea habitat

bunk room

dining area

toilet

wet porch

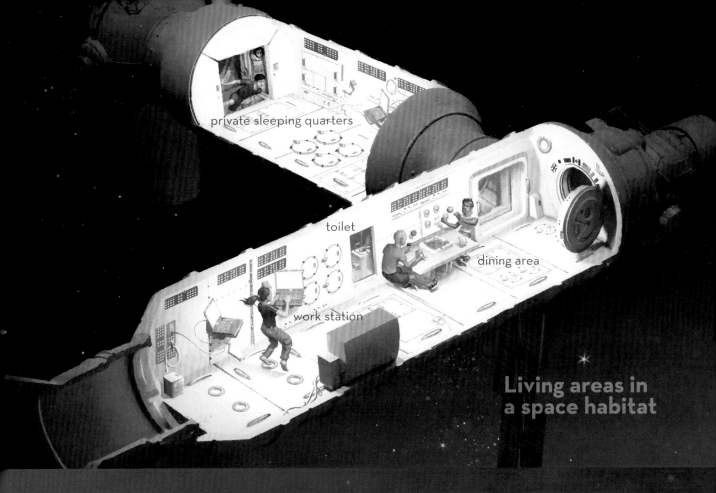

private sleeping quarters

toilet

dining area

work station

Living areas in a space habitat

You've made it. You're here! You may be floating gently through the International Space Station (ISS). Or perhaps you are safely settled inside your underwater research center. Wherever you are, you are set for a long-term stay. Astronauts are typically in space for an average of six months. After all, it took you more than two years of training to get there, so you may as well stay awhile. Aquanauts stay underwater anywhere from one to two weeks, but usually no longer than a month at most. It depends on the size of the research station and the length of their project.

LIVING SPACES

The experience of living in space and deep under the water is very similar. Both places must have their own supply of oxygen, food and water, and toilets. They must have living spaces, working spaces, and access to communication with the outside world. They must have safety equipment for medical emergencies and ways for the astronauts or aquanauts to get out quickly if necessary. They need air locks or wet rooms for transition from the outside to the inside environment. They need to have their own power systems for lights, electricity, and air. In addition, these craft must be secure from the environment outside. You don't want a hole in either your spaceship or your submersible. Exposure to space would immediately suck all of the oxygen out of the spaceship. A tiny hole in an underwater station would cause water to pour inside.

> > >

UP: LIVING IN SPACE

The ISS is currently the only long-term dwelling in space. It is about the size of a five-bedroom house and can safely hold a crew of six people. The station can accommodate up to 10 people living and working there. The extra number is for when astronauts or cosmonauts, Russian astronauts, fly up to join the crew or are getting ready to leave the ISS to head home. The ISS has living areas, research labs, sleeping areas, and of course, an air lock for docking spacecraft. It's a combined effort of five different space agencies from the United States, Russia, Japan, Canada, and Europe. Each agency has its own research lab and living quarters. There is a joint area for advanced life support, oxygen generators, water recycling, a treadmill, and a toilet.

DOWN: LIVING IN THE OCEAN

If you are chosen to do research under the sea, you will probably be going to Aquarius. Aquarius is one of the only functioning long-term underwater research stations in existence. It is located 60 feet (18 m) below the surface in the Florida Keys National Marine Sanctuary. Aquarius sits on a sand patch and is next to a coral reef, which is a great place for research and study. The Aquarius research station is an 80-ton (72-t) cylindrical chamber. If it were on land, it would weigh about as much as 10 elephants. Aquarius is much smaller than the ISS. It is only 43 feet (13 m) long and 9 feet (2.7 m) wide. That's about the size of two grown elephants standing in a row. There is room for four scientists and two technicians to live onboard. The main living area, or main lock, contains six bunks for sleeping, computer work stations, two large windows, and a kitchen. The entry lock is a smaller room where you will find communication equipment, the air conditioning and heater controls, a sink, a toilet, and life-support equipment. An important part of Aquarius is the wet porch. It is how the crew

ASTRONAUT TRACY CALDWELL DYSON LOOKS THROUGH A WINDOW IN THE CUPOLA OF THE INTERNATIONAL SPACE STATION.

What is it like to LIVE IN SPACE for a LONG TIME?

It is so wonderful to live in space—on a space station—that I didn't really want to go home! In fact, very quickly, it felt like home to me. I loved feeling like a pioneer, getting to explore our solar system and come back and tell everyone all about it. I loved looking at planet Earth and realizing that everyone I knew was down there, and I was here, on a space ship. I realized that we are all from one place, and that the borders between countries don't separate us as people.
—Cady Coleman, astronaut

members enter and leave the research station. Unlike the ISS, where the doors are on the sides, Aquarius has a door on the bottom or floor of its station.

In order to get into Aquarius, you will scuba dive down to the wet porch and come up from the sea access point. The wet porch is about the size of a normal bedroom. It is fully enclosed and has an air lock at one end that connects it to the rest of the living space. The air pressure can be increased or decreased depending on what is needed for the divers. This is where you will store your scuba gear, along with the rest of the crew's gear. Next to the wet porch is a gazebo. It is there for emergencies in case of total power failure on Aquarius. It has enough air for six crew members to stand and quickly don their scuba gear for a trip to the surface.

What keeps Aquarius in position? The entire station is anchored to the seafloor by a 116-ton (105-t) baseplate. Each of the four legs that support the station is filled with 25 tons (22.7 t) of lead. The legs are sunk into the ocean floor for stability. They can be adjusted up to seven feet (2.1 m) up or down to keep Aquarius level. That's a good thing, since sometimes the huge storms that roll across the water have deep waves that can affect the station.

Three weeks is a long time to live underwater. Aquanauts can be subject to decompression sickness from the change of pressure. To prevent that, aquanauts in Aquarius work at what is called saturation diving. The pressure in Aquarius is kept at about 2.5 times atmospheric pressure. A constant pressure is needed for the aquanauts to make their daily dives. So, the atmosphere within Aquarius is at a slightly higher level than people feel on Earth's surface. When the aquanauts arrive at Aquarius, they fairly quickly adjust to the increase in the natural atmosphere, and they begin to move about freely and work normally. Being exposed to the same pressure inside as outside the habitat allows them to be able to dive for longer periods of time. Their bodies have adjusted to the extra nitrogen in their blood.

Does that make them feel different? Some aquanauts have noticed that their ears feel pressure, or that their chest might be a little tight. Some feel a little sick, like being carsick. Others don't notice anything at all. When diving from Aquarius, aquanauts are required to stay at depths of 47 feet (14 m) or lower at all times to remain in saturation diving. This means that when they are outside of Aquarius, they have to be very aware of their depth while they are swimming around. If they go too high, they will experience decompression issues.

At the end of the mission, the pressure within Aquarius is slowly reduced over a 17-hour period. Then the divers move to the wet deck where the pressure is the same as on the surface. Finally, they slowly ascend to the surface.

What is it like to LIVE UNDERWATER for a LONG TIME?

One of the coolest things about living underwater is how quickly the undersea habitat and the reef around it start to feel like home. You begin seeing the same animals every day—an enormous fish called a grouper that lives under the habitat, and a barracuda trailing a fishing lure in his mouth. During one of the missions, an octopus laid her eggs under the window hatch next to the tiny table where we ate dinner every night. At night you go to sleep in your bunk to the sound of thousands of snapping shrimp clicking away on the outside of the hull.

—Brian Helmuth, aquanaut

SATURATION DIVING

In space the most important thing is to keep the pressure in. Space has no atmosphere, and without pressure, the human body would expand dangerously. Underwater the most important thing is to keep the pressure out. Atmospheric pressure, the amount of pressure we feel just walking around on Earth, increases as you go deeper. Scuba divers must always be aware of how deep they are swimming, and how long they are underwater. That's because higher pressure makes the human body change how it works. When we breathe on land, our bodies absorb oxygen (as a gas) from the air. The oxygen goes into our bloodstream and is carried to our lungs. Our lungs move in and out to take in more oxygen and release carbon dioxide.

When a scuba diver goes below 33 feet (10 m), the pressure on his or her body actually doubles. This dramatic increase in pressure makes the lungs contract. Basically, it would feel as if someone grabbed your lungs and squeezed them like you'd squeeze a sponge. This makes it harder to breathe and take in oxygen. It also feels pretty uncomfortable. Of course, that's why you wear the air tanks. The tanks provide pressurized air that allows your lungs to relax and you to breathe normally. Scuba diving tanks have air, but saturation diving—staying below 33 feet (10 m) or deeper for long periods of time—requires a different type of air. Instead of using normal air to breathe, saturation divers who dive very deep breathe a mixture of gases: oxygen, nitrogen, and heliox. Heliox is a mixture of 79 percent helium and 21 percent oxygen. Unlike nitrogen, which goes into the tissues of scuba divers, heliox is handled much better by the body. Heliox is safer and easier for saturation divers to breathe and is also eliminated from the tissues much more quickly than nitrogen.

Why go down that deep if it changes how your body works? Unlike surface diving, where we can at most spend about two hours per day at 60 feet (18 m), saturated divers can work for periods of nine hours or more per day at that depth. Essentially, being saturated moves the "surface" down to 47 feet (14 m).

The process for getting back to the surface for both scuba and saturated divers is similar. While they were diving, their bodies took in enough extra gases to allow them to be at equilibrium with the surrounding pressure environment. These gases must be expelled before the divers return to the surface. This is why both divers have timed stops on their ascent.

If the divers rise too quickly, the nitrogen will form bubbles in their blood and burst. It's like when you open the top of a soda can. The bubbles instantly rise to the surface, causing fizz. While that may be good for the soda, it's definitely not good for you.

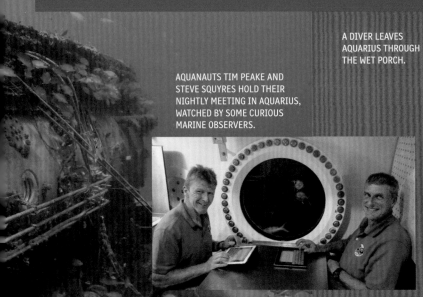

AQUANAUTS TIM PEAKE AND STEVE SQUYRES HOLD THEIR NIGHTLY MEETING IN AQUARIUS, WATCHED BY SOME CURIOUS MARINE OBSERVERS.

A DIVER LEAVES AQUARIUS THROUGH THE WET PORCH.

LET'S GET
TO WORK

Time spent in either space or the deep sea is precious, since it is usually short. The astronauts' and aquanauts' primary mission is to conduct research and keep up the maintenance on their stations.

UP: WORKING IN SPACE

During your work in the morning, you might be checking on some cool experiments. Ones that have made it to the space station include growing zucchini, or watching how ants move around as a colony in microgravity. You may be trying out the new 3-D printer or, if you're lucky, be in charge of the maneuvering Robonaut for the day.

One of your main jobs will be to keep the ISS running in tip-top shape. Keeping the parts working both inside and outside the station is very important. It's kind of hard to get a plumber or an electrician up to the space station to fix something. You and the other astronauts must fix everything yourselves. Occasionally, that means that you will have to put on your space suit and go for a walk. You might need to fix the solar power system or remove space "junk" that is stuck to the outside of the station. Fixing things sometimes requires creativity. After all, you can only use the parts that are already onboard. Astronaut Sunita Williams once had to use her own toothbrush to fix the solar power system on a space walk. Don't worry. She had a spare.

Meet
ROBONAUT—
Astronaut Helper
EXTRAORDINAIRE

>>

AT A SPACE-HABITAT SIMULATOR IN THE SOUTHWEST-ERN U.S. DESERT, DR. ROBERT HOWARD CHECKS ON AN EXPERIMENT FOR GROWING VEGETABLES IN SPACE.

A RESEARCH DIVER RECORDS OBSERVATIONS ON AN UNDERWATER CLIPBOARD.

DOWN: WORKING UNDERWATER

One of your main jobs as an aquanaut is to study the environment around you. More than likely your mission will be for a specific time in which to conduct prearranged research. Perhaps you want to learn more about the coral reef surrounding the research station, like how it survives and thrives in its marine environment. Or maybe you want to study the aquatic animals that live in and around the reef.

Whatever your research, you will be conducting it by making dives, going out into the water, and coming back in. Then you will write up your notes and observations for further study when you get back on land.

You may be asked to do maintenance on Aquarius. After all, the crew who live onboard are the ones who need to keep things running. If you are helping train future astronauts, you might be showing them how to maneuver large objects in a fluid environment, similar to what they will have to do in space.

NASA thought that the astronauts could use an extra set of hands, so they created Robonaut. It is a fully-automatic robot that can "think" for itself. Its software programming is actually what does the thinking. Astronauts give it small tasks, such as using a screwdriver or writing notes. Robonaut figures out how to do these things by itself. It has a head, body, arms, and hands that work much like a human's. Robonaut can hold tools and even type on a computer keyboard. In fact, Robonaut sent its first tweet in July 2010. Robonaut can be attached to a platform with wheels so that it can move. It has recently gotten legs. Future plans may even include a solo space walk. Talk about out-of-this-world robotics!

COMMUNICATION

Both astronauts and aquanauts are in constant communication with people on Earth. The main reasons are to exchange information and to keep the astronauts safe. Astronauts are required to report in at least twice a day but can be in contact with the Earth at any time. If any medical issues or problems with the electronic systems or space walks arise, the astronauts immediately inform Mission Control. The engineers on the ground are trained to give expert help and guidance to the astronauts. Open communication between astronauts and Mission Control hasn't always been possible. It used to be that Mission Control could only talk to the orbiting spacecraft when it was within a certain distance from the building. Now the Tracking and Data Relay Satellite system is in place. It consists of seven operational satellites that track the ISS and other spacecraft and maintain a constant open line of communication with Mission Control on the ground.

It's easier for aquanauts to maintain open lines of communication, as Aquarius is tethered to an antenna buoy that can send information back to shore-support base. Aquarius has a Mission Control center just like the astronauts do. There is an umbilical cord, or long line, that connects Aquarius with the surface. This allows for radio and data communication and for supplies to be sent down if needed. Its most important job is to provide oxygen and air supply. Manned submersibles are also in constant contact with their research ships. This is important for safety and also as a way to transmit data to the researchers onboard the ship.

TOOLS AND TECHNOLOGY

Astronauts and aquanauts have many tools to help them in their work or research. Some are normal tools you'd see in a toolbox, such as a screwdriver, drill, or hammer. Others are much more complex and help the explorers go places they couldn't reach safely on their own.

UP IN SPACE

The ISS has a robotic arm called the Canadarm 2. It got its name because the Canadian government paid $1.1 billion to help create it. The 57.7-foot (17.5-m) arm has seven different motorized joints. These joints allow it to move in many different directions and reach a lot of awkward spaces. It is controlled from inside the ISS and is attached to a mobile system of rails along the outside of the ISS that allow the astronaut "driver" to move the arm wherever it is needed. Plus, it can lift and move all kinds of heavy objects. The arm can carry more than 200,000 pounds (90,000 kg) of heavy cargo. Or it can gently pick up an astronaut and transport her from one place to another.

Another tool used regularly by astronauts is the pistol-grip hand drill. It is a specially designed drill that can be used with the heavy space gloves an astronaut wears. Sometimes astronauts need to make repairs to satellites or even the ISS itself. This handheld drill makes things much easier. Still, the astronauts' most important tool might actually be the 25-foot (7.6-m), heat-resistant tether. This remains attached to an astronaut's belt whenever he or she ventures outside of the ISS. Its job? To keep the astronaut from floating off into space. A very important tool indeed.

AN ARTIST'S DEPICTION OF THE EUROPEAN ROBOTIC ARM ATTACHED TO THE INTERNATIONAL SPACE STATION

SPACE ARM
Dives Deep

A new piece of equipment headed to the International Space Station took a dip in the waters down below first. The European Robotic Arm (ERA) was tested underwater by several Russian cosmonauts. Their mission was to take it apart and put it back together and to test its moving capabilities. This is to ensure that the ERA can be successfully assembled in space and that it can be disassembled and put back together in case of repair. It made sense to test this out underwater because the ERA is more than 36 feet (11 m) long and weighs almost 1,400 pounds (630 kg). That would make maneuvering it on land practically impossible.

>> DOWN IN THE DEEP SEA

Aquarius is equipped with a simulated robotic arm similar to the ones used in space. It allows astronauts to train when they are underwater. Almost all space conditions can be mimicked deep under the ocean. While communication is relatively quick on Aquarius, the responders purposely wait the 10 to 15 minutes it can sometimes take in deep space for a response. Astronauts-in-training are put through real-time tasks that they would perform in space. They must move the robotic arm, using it as a crane for large objects or possibly each other. They also learn to maneuver other vehicles such as a lander or rover on the bottom of the ocean next to Aquarius.

The one thing that Aquarius has that space does not is an autonomous underwater vehicle (AUV). These are unmanned submersibles. That means they don't have people inside. They are programmed to go off on their own to collect data and then return. The AUVs can travel much farther than a human could in a safe travel time.

These can be used to take videos, map complex environments, and explore other habitats. New AUVs are being designed to hover next to aquanauts and support them in their tasks. The AUV might deliver tools to an aquanaut or recover objects that might have been dropped or slipped from the aquanaut's grasp. The idea behind the development of these AUVs is for them to be used in space one day—perhaps on a trip to Mars.

NASA ASTRONAUTS ON THE OCEAN FLOOR PRACTICE MANEUVERS THEY WILL NEED IN SPACE.

NASA ASTRONAUT KATE RUBINS PARTICIPATES IN AN EMERGENCY DRILL ABOARD THE INTERNATIONAL SPACE STATION.

TYPICAL DAY
IN SPACE

Your schedule in space would look a lot like your schedule when you're in school. Except of course you wouldn't have to leave your house to go to school or, in this case, work. You wake up around 7 a.m. and float to the main room for your morning conference at 7:10 a.m. Next you eat breakfast and maybe quickly brush your teeth. You don't worry much about combing your hair. It's just going to stick up anyway. From 8 a.m. until noon you are in the lab, working on research or maybe checking on experiments that were brought onboard. You stop for 30 minutes for a quick lunch and then you are back working until 6 p.m. You're not done yet. You have another conference with everyone to end the day. Finally, you get dinner. From 7:30 p.m. to midnight you are on your own. You might choose to connect with family, work on things for the next day, or maybe just stare out the window and watch one of the 16 sunrises and sunsets that pass by each day. Perhaps you will do your required exercise at night. Every astronaut must do 30 minutes on the treadmill and 70 minutes of resistance training every day. (Think exercising with a giant rubber band.) This keeps their muscles and lungs in shape.

Astronauts work pretty much every weekday, but they do get weekends off. They also get together on Fridays to watch television. Movies and television shows are beamed from stations on Earth. The television connections are not always the best, though. After all, the space station is moving so fast that it orbits the Earth every 90 minutes. That speed can sometimes make the picture freeze or get blurry.

TYPICAL DAY IN
THE DEEP SEA

Your typical day in Aquarius may start with a morning briefing and breakfast. Dives are planned in advance and must be approved by the Aquarius mission coordinator. You will sit down with the crew members who are going on the dive and review exactly what will happen. Then you'll check your scuba gear. You need to make sure that it has the proper pressures in the air tanks and that it is full.

Next, you'll be partnered with a buddy. There is a strict buddy system for all dives. This is to make sure that if one person gets into trouble during the dive, there will be another to help him or her get back to Aquarius. During the briefing you will go over specifics of where you will be located during the dive. Most important, you will need to understand how long you are able to dive. Time limits are strictly enforced for each dive. It is a matter of safety that each person be diving for only a certain amount of time. Like spacewalking, diving affects the body. Staying out too long in space or underwater can have very negative effects on a person.

Once your briefing is complete, you will head to the wet deck and put on your scuba suit. When your gear is set, you and your buddy will proceed through the sea access point.

Ughh—shhhhhh. You are now breathing through the respirator on your scuba gear. You blink as your eyes adjust to the cool darkness. The world comes into focus. The amazing and unique underwater world stretches before you. A school of fish envelope and surround you on their way to the reef. With a flip of your fins you glide through the water, off to explore the ocean bottom.

You can spend from three to six hours diving. Then you will need to return to Aquarius for at least four hours of rest. During that time you may write up notes and have something to eat. You are allowed one more dive of no longer than three hours in a one-day period. After your two dives of the day, you can do what you want. You might write up notes, communicate with your family, rest, or get some much needed sleep. Aquanauts' favorite pastime is to stare out the giant window. They enjoy watching the many different types of aquatic creatures swim by. At night they sometimes shine a flashlight out the window to attract zooplankton, microscopic aquatic animals. Shortly after, the small fish and even barracuda come near to eat the plankton. Who needs TV? Aquanauts have their very own aquarium right outside their window.

AFTER YOUR MISSION

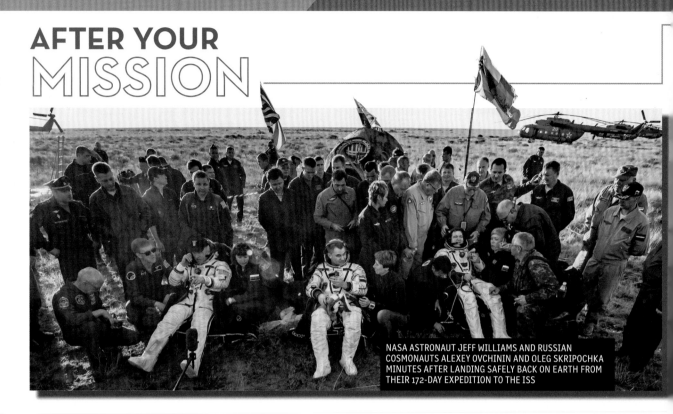

NASA ASTRONAUT JEFF WILLIAMS AND RUSSIAN COSMONAUTS ALEXEY OVCHININ AND OLEG SKRIPOCHKA MINUTES AFTER LANDING SAFELY BACK ON EARTH FROM THEIR 172-DAY EXPEDITION TO THE ISS

ASTRONAUT

Upon returning from space, astronauts will experience many changes to their bodies. First of all, it may be tough for them to walk. Your muscles will have lost some of their strength since, without gravity, you haven't had to use them in quite some time. You will have lost calcium, which is important for bone strength. That means your bones may be more brittle and break easily. Your heart is a muscle and hasn't had to work as hard in space, so you may find yourself out of shape and out of breath.

On Earth, gravity keeps the blood flowing downward, but in microgravity, it flows upward. Because of this, the blood in your body goes to your head. This can cause higher pressure in the eyes and affect vision. You may have rosy red cheeks from increased blood flow, and maybe even a headache. You may feel a little off balance and even dizzy. During the space flight, the fluid

HOW DO YOU FEEL AFTER YOU COME BACK... TO EARTH?

On my four short space shuttle missions, there was some feeling of heaviness and balance problems on landing day, but after that first day back, I experienced no effects of being in weightlessness. On the 115-day Russian mission, I had lost significant muscle and bone mass, my heart had gotten weaker, I was anemic, and had balance problems for several days after landing. Some of those conditions took weeks to correct, and I never recovered all of the lost bone mass.
—Norman Thagard, astronaut

in your ears might have sloshed around. Now that you are on Earth, your ears need to readjust to the constant atmospheric pressure.

Don't worry. All of these issues are normal and go away rather quickly. In most cases, they can be fixed by resting and slowly resuming exercise.

AQUANAUT

After spending three weeks in Aquarius, the aquanauts are required to undergo a 12-hour saturation observation. This takes place in the base station on the surface. Aquanauts who are under saturation observation are checked over physically and watched for any sign of ill effects from their long-term stay in Aquarius. They are encouraged to rest and recuperate as their bodies readjust to the surface's normal pressure. Once they're cleared to go, aquanauts gradually assume their normal activities. For the next few days they are cautioned to avoid any strenuous activities like running or other exercising. Since their bodies are still getting used to higher oxygen concentrations, using up a lot of oxygen while exercising is not a good idea. They should also avoid taking hot showers, as this could upset their body compressions. Finally, they are not allowed to fly for at least 48 hours, or two days, after decompressing. Going up in an airplane reduces the pressure on the body. This could allow nitrogen to seep back into their tissues. They would have the same decompression problems as the plane descends. That is something to be avoided until their oxygen-rich blood system has stabilized.

Going into space or deep underwater for long periods of time can be tough on a person. It's demanding work that affects your body, your mind, and even your family life. So why would people want to do this? The love of exploration and the urge to learn more about the world we live in!

TO THE SURFACE?

Returning back to the surface at the end of the mission feels very strange, as if you belong underwater and are only visiting the land. Everything seems noisy and very bright, and with far too many people. Physically I've always felt fine after missions, although after spending up to nine hours per day in the water your skin and ears really take a beating. Still, I was ready to go back down in a heartbeat.
—Brian Helmuth, aquanaut

EXPLORER'S NOTEBOOK

3 LIVING AND WORKING ENVIRONMENTS

✓ Allow your body to adjust to your new "home."

✓ Keep your equipment in good working order.

✓ Exercise and get a lot of sleep.

✓ Be aware of the forces acting on your body.

✓ Don't forget to relax and have fun.

CHAPTER 4 WHY DO WE EXPLORE?

1 B.C.   1700s   1900s   2016

1609   1969   1997   2015

Humans are curious. We strive to understand things, to solve mysteries, and to discover new places. Ancient astronomers such as Galileo and Copernicus looked to the stars for help in understanding how the Earth fits into the universe. Deep-sea explorers William Beebe and Jacques Cousteau spent thousands of hours underwater seeking to understand the ocean environment. To fulfill our need to explore, we have spread across all of the continents, and sent satellites and probes to the farthest reaches of space and the deepest depths of the sea. We have made discoveries that have changed the way we view our planet. That is why exploration is so important. It gives us knowledge of our own planet so that we can understand it and, hopefully, take better care of it.

This is what you will be doing if you become an astronaut or aquanaut. You will train to go into space or dive deep, or perhaps you will learn how to control a remotely operated vehicle (ROV) from inside a safe spacecraft or ship. Either way, you will be one of the amazing explorers tasked with bringing a new understanding of how humans fit into the universe.

EARLY EXPLORATION

One of the biggest parts of exploration is gathering data. Early explorers were unable to travel to space. Instead, they made telescopes to observe the skies. They made maps of what they saw and came up with ideas for how the universe worked. Voyagers of the oceans did the same. They traveled on ships and made maps of the seas to navigate, or find their way. The ocean was within reach. Salvage companies were diving underwater to retrieve items from sunken ships even before 1 B.C. and into modern times.

INVENTOR JOSEPH PERESS SHOWS OFF HIS
NEWLY DESIGNED STEEL DIVING SUIT IN 1925.

At the time, the divers just held their breath and dove. Later, early versions of a diving helmet were developed. The diving helmet was simply a bell-shaped metal container worn over the head. Air was trapped in the bell as it was submerged. The person could breathe as long as the air stayed in the bell. The problem was that the air did not last for very long, and the person was not able to dive very deep. It wasn't until the 1700s that better, more efficient diving bells were created. These allowed divers to go as deep as 36 feet (11 m). This seemed to be the point where diving became difficult. At this depth, the pressure on the human body is twice that of atmospheric pressure. Without a pressure suit, the divers' lungs would constrict and their ears

AN INSTRUCTOR CHECKS ON A STUDENT LYING IN A DECOMPRESSION CHAMBER DURING A 1930 DIVING CLASS.

would hurt. One diver explained the pressure in his ears as feeling like "a quill [or pen] was being stabbed into each ear." Ouch!

The first all-in-one diving suits with pressure control were designed in the early 1900s. They

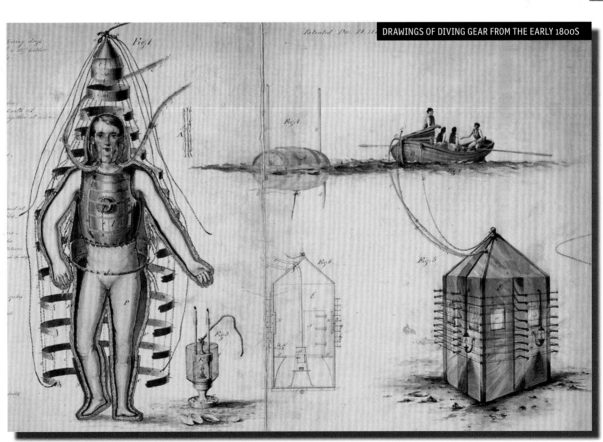

DRAWINGS OF DIVING GEAR FROM THE EARLY 1800S

were made of metal and extremely heavy. A suit made of aluminum alloy weighed 550 pounds (250 kg). That's about as much as a male African lion. Can you imagine wearing that? You wouldn't be able to walk! But in the water, the diver could go over 200 feet (60 m) deep because the suit held up to four hours of air.

How does this relate to the advances we've made in space? You may notice that the early diving suits look a lot like the space suits that the astronauts wear today. In the early 1960s, when people turned their sights to exploring space, they used the diving suits as a guide for how to create space suits. The first "astronauts" were fruit flies. They were sent up in 1947 to see how space would affect their bodies. Why a fruit fly? They have many genes similar to those of humans. Genes are bits of information that determine characteristics that you get from your parents, like eye color, hair color, the length of your nose, and how tall you are. Genes also help determine what kind of diseases you might get. That's where the fruit flies come in. They share almost 61 percent of all the known disease genes with humans. The NASA scientists figured

The **DEEP SPACE** Suit/ The **FUTURE** Suit

As we look to take astronauts to Mars, the suits they wear must be able to fit comfortably, keep the astronaut under constant atmospheric pressure, and control temperature. They will have a protective shell and self-healing gloves to fix any tears that happen immediately. They may even have their own gravity systems so that astronauts' muscles don't diminish.

The Huntsville Times

Man Enters Space

'So Close, Yet So Far,' Sighs Cape

U. S. Had Hoped For Own Launch

Soviet Officer Orbits Globe In 5-Ton Ship

Maximum Height Reached Reported As 188 Miles

that if the fruit flies could survive space, so could humans. Besides, fruit flies are small. NASA could use a tiny spaceship, and the fruit flies only needed a few kernels of corn for food.

Still, the goal was to send humans to space. That happened in 1961 when the first cosmonaut, Yuri Gagarin, completed one Earth orbit and landed two hours later. Eight years later, when Neil Armstrong and Buzz Aldrin touched down, the United States became the first country to land humans on the moon. Today, technology has advanced so much that both diving and space suits allow us to go farther and deeper than ever before.

A DIVER IS LOWERED INTO THE WATER TO RECOVER OBJECTS FROM A SUNKEN SHIPWRECK OFF THE COAST OF CHILE AROUND 1900.

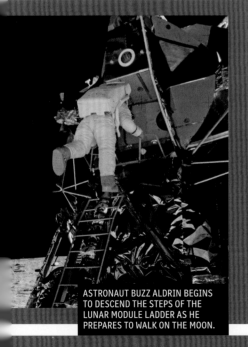

ASTRONAUT BUZZ ALDRIN BEGINS TO DESCEND THE STEPS OF THE LUNAR MODULE LADDER AS HE PREPARES TO WALK ON THE MOON.

When you are designing a space suit, it is really important to think about what it will be like for the astronauts that are wearing the space suit and how they will be able to do all their different tasks while they are wearing it. But because there is a lot more gravity on Earth than on the moon, Mars, or an asteroid, space suits feel very heavy to astronauts when they are testing them on Earth—similar to trying to walk around with a really heavy backpack. So to make the testing more realistic, we can take the space suits underwater and then add just the right amount of floats to the space suits to make it feel as though there is less gravity. Then you can pretend to float across an asteroid or bounce around on the moon, depending on how many floats are added. It is important testing, but it's also a lot of fun!

—Andrew Abercromby, NASA biomedical engineer, who has worked on developing the Extravehicular Mobility Unit

THE NASA SPACE SUIT/EXTRAVEHICULAR MOBILITY UNIT

With 14 layers and 1,800 moving parts, a space suit provides everything an astronaut might need on a space walk. Pressure and temperature control, oxygen, and radiation protection are all offered in this two-piece semirigid suit.

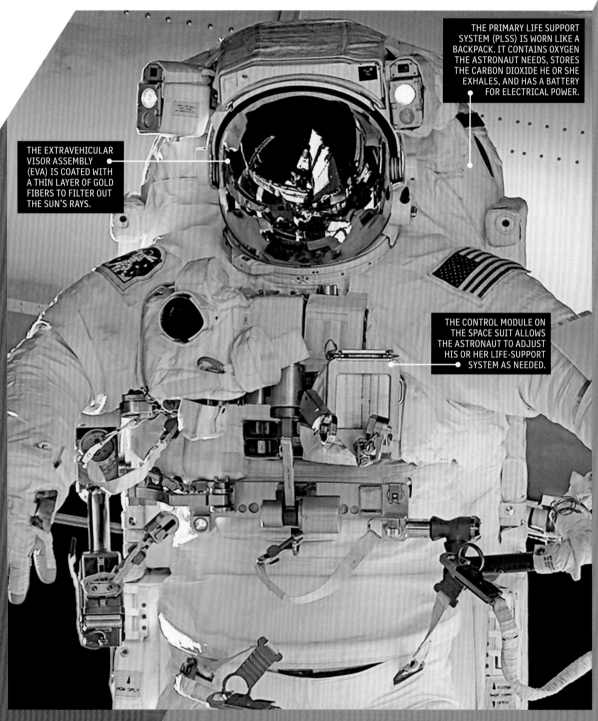

THE PRIMARY LIFE SUPPORT SYSTEM (PLSS) IS WORN LIKE A BACKPACK. IT CONTAINS OXYGEN THE ASTRONAUT NEEDS, STORES THE CARBON DIOXIDE HE OR SHE EXHALES, AND HAS A BATTERY FOR ELECTRICAL POWER.

THE EXTRAVEHICULAR VISOR ASSEMBLY (EVA) IS COATED WITH A THIN LAYER OF GOLD FIBERS TO FILTER OUT THE SUN'S RAYS.

THE CONTROL MODULE ON THE SPACE SUIT ALLOWS THE ASTRONAUT TO ADJUST HIS OR HER LIFE-SUPPORT SYSTEM AS NEEDED.

A CAMERA OPERATED BY AN AQUANAUT RECORDS IMAGES UNDERWATER.

18 ROTARY JOINTS GIVE A DIVER GREAT MANEUVERABILITY UNDERWATER.

TWO CLAWLIKE GRABBERS ALLOW THE AQUANAUT TO GRASP OBJECTS UNDERWATER.

THE EXOSUIT

The Exosuit, a newly designed diving suit, can take an aquanaut to 1,000 feet (305 m). The suit keeps your body at atmospheric pressure so that you don't have to worry about decompressing as you come back up.

BEYOND HUMANS

While it is great to have astronauts and aquanauts, exploration does not always require human participation. In fact, it's not always practical. Human exploration has limits—after all, humans need to have certain temperatures and pressures to survive, not to mention oxygen. That isn't found in every place that we want to explore. At present, technology allows us to go only so far in space or deep underwater. We must rely on other ways to gather information. That's why we use unmanned space probes, satellites, and deep-sea, remotely operated vehicles.

TECHNOLOGY ALLOWS US TO GO ONLY SO FAR IN SPACE OR DEEP UNDERWATER.

Just as the suits used to explore space and the deep sea are similar, the unmanned vehicles look somewhat alike. They certainly have the same goal: to explore new environments without risk to human life. These unmanned craft are packed full of sensors, computers, long-distance communication systems, and cameras. They are designed to be operated remotely by humans or programmed to travel specific paths. Let's take a closer look at two unmanned ROVs to see how they work.

DIFFERENCES:

The Cassini-Huygens is larger and has a huge satellite dish. It also has heat shields and thermoelectric generators. This is to make use of the solar power readily available in space.

The Jason/Medea ROV is more compact and rounded. This is to reduce the drag from the water. Larger objects move more slowly in water. The curves help the water flow more easily across its surface.

The Cassini-Huygens was designed to take seven years to reach Titan, one of Saturn's moons. It will not return.

The Jason/Medea ROV is deployed for two to seven days for short research trips to the depth of the ocean. It can be retrieved by the support ship at any time.

The Jason/Medea ROV is attached by cable to the support ship floating on the surface.

AN ARTIST'S DRAWING OF THE CASSINI-HUYGENS CRAFT

THE CASSINI-HUYGENS IS CONTROLLED BY SCIENTISTS BACK ON EARTH.

JASON/MEDEA ROV

SIMILARITIES:

Both have a two-body system. One part of the vehicle is for gathering information. The other part is for stabilizing and providing power.

Each of these vehicles is equipped with its own propulsion system.

They both have navigation systems and are controlled remotely by humans.

They both have emergency backup systems, video, communication lines, and cameras.

JASON

WOODS HOLE OCEANOGRAPHIC INSTITUTION

JASON NOAA ONR

The **Cassini-Huygens** craft is broken into **two parts.** The Cassini is the orbiter; **Huygens is the probe** sent to the surface of Titan.

HOW THEY WORK

In the *Jason-Medea* ROV, the *Jason* part contains sensors, video equipment, cameras, and lighting systems. It has a robotic arm that can capture samples of sediments, bits of rock or sand, larger rocks, or even marine life. The arm places the samples into a basket, or it can set them on a larger platform that elevates them to the surface. *Jason* carries a sonar system, which allows it to map the area it is going over. *Medea*'s job is to keep *Jason* stable so that it can get a good video image. It also provides overhead light so that as much of the seafloor is lit as possible. That is important because the depths at which the ROV operates are pitch black.

The *Jason/Medea* ROV is attached via a long cable to the support ship floating on the surface. This is where the scientists control the ROV. They are able to see real-time video images that are sent back from deep underwater.

The Cassini-Huygens craft is also in two parts. The Cassini is the orbiter, which means it provides the navigation for the spacecraft. It also contains sensors that it uses to record information while in orbit. This gives the scientists a sort of bird's-eye view of everything. Cassini is directed by a navigation team of scientists and engineers here on Earth.

Once the Cassini orbiter reached Saturn's largest moon, Titan, the Huygens probe was released. It parachuted to Titan to investigate its surface. Yet, the probe still remained connected to Cassini via the probe-support equipment (PSE). Huygens's many different sensors sent information to Cassini on the makeup of the atmosphere, the temperatures, and the electromagnetic waves it encountered.

SHIPS AND AUVs

The deep sea has a few advantages for exploration that space does not. It is closer. It is much easier for humans to access. Your mission as an ocean explorer may be onboard one of the several research ships that are in use by the National Oceanic and Atmospheric Administration (NOAA). These ships are basically floating laboratories equipped with special tools and technology to explore the world's oceans.

You may be one of the crew in charge of directing the sonar to scope across the ocean floor. Or perhaps you will be asked to pilot one of the ROVs as it descends to great depths. Your job may be to gather information from the onboard sensors and send it back to ground stations for more scientists to study. Either way, you will live onboard a ship for weeks or months, depending on the length of its mission. Each ship has all the supplies its crews need for months at sea. You may travel to any of the five oceans, into the polar regions, or possibly even over the Mariana Trench, the deepest part of the ocean. You will learn about water temperatures, pressures, currents, and climates. You may even come face-to-face via video with some of the most amazing and unique creatures of the deep. Sound like fun? Absolutely! Important work isn't always fun work, but the job you will be doing on this research vessel will help us learn more about the biggest part of our planet: its oceans.

Another way that you might be asked to help explore the oceans is by managing an autonomous underwater vehicle (AUV). These are computer-controlled, unmanned vehicles that move through the ocean on their own. They are not tethered to anything but are guided by their onboard computer. AUVs can be smaller than 100 pounds (45 kg) or weigh more than 1,000 pounds (454 kg). Some are solar powered and must come near the surface to reach the sunlight for energy. Others have batteries that can run for months.

The mission of an AUV is to collect data wherever it goes. AUVs can move with ease from shallow waters to very deep waters. They can go where ships and larger submersibles cannot. Information from an AUV is transmitted via acoustics, or sound energy, to research stations. While the AUV "decides" where it goes most of the time, the operator can give short, specific commands like "stop" and "return home."

What are AUVs used for? To locate underwater obstacles (like shipwrecks), live underwater mines, and even locations of large fish colonies for fisheries. They are also used to map the ocean floor and coastal regions. The information gathered by AUVs helps us understand weather patterns, water quality, and even the effects of climate change on the ocean.

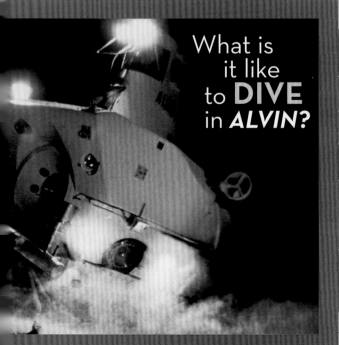

What is it like to DIVE in *ALVIN?*

We slowly descended through the water, passing through the layers of diminishing light and into the darkness. Once in the complete dark, we were treated to an underwater fireworks show as bioluminescent animals danced and crashed into each other, leaving 'sparks' of luminescent material in the process. Upon reaching the ocean floor, the pilot steered *Alvin* through the black depths, in and around lava pillars and sulfide spires created by deep-sea hot spring or hydrothermal activity. We knew that few others had ever visited the site that we were diving on, and, due to the dynamic nature of the ocean floor, we were prepared for any and all surprises.

—Dawn Wright, aquanaut

EXPANDING OUR
HORIZONS

Over the past 60 years, the space program has sent hundreds of probes into space, from the Mariner spacecraft, which mapped the surface of Mars in the 1970s, to the New Horizons space probe, which sent back close-up images of Pluto in 2015. The amount of information collected from these unmanned spacecraft has been very helpful to scientists. It has increased our knowledge of many of the planets, asteroids, and moons in our solar system. And yet, there is still so much more to learn. New probes and spacecraft are being planned and designed for future trips. NASA is focused on developing a way for humans to land on Mars. It is also interested in deep-space communication and learning how to grow food crops in space. That would solve the problem of having to resupply the ISS with foodstuffs.

Deep-sea exploration has been happening for almost twice as long as space exploration, but the advances have been much slower. Why? To be blunt, we love space. We watch space movies, we read space comics, and we watch astronauts floating in space. We cheer with the scientists when we see the first images from Pluto. Let's face it. Space exploration is exciting!

And yet, deep-sea exploration is just as awesome. The oceans cover more than 70 percent of our planet. Their vast, alien environment is home to millions of creatures and great wonders that we have only just begun to unearth. The oceans are a part of our planet, and we may need to look to them in the future for food, energy, and even changes to our climate. We already use the ocean for food and water. We have harnessed the waves for energy and used waterfalls for hydroelectric power.

FRESH FOOD
in SPACE?

RED ROMAINE LETTUCE PLANTS GROW INSIDE A PROTOTYPE VEGGIE FLIGHT PILLOW.

You take it for granted. When you're hungry you can probably go to the refrigerator and just grab an apple or make yourself a fresh salad. If you are thousands of miles in space you can't do that. Astronauts must live on prepackaged foods and drinks. Eating and drinking that for months on end can get kind of old. Scientists are trying to fix that. Special plant growth chambers have been installed on the ISS. These "plant pillows" are being used to grow romaine lettuce and radishes under special lights. The result? So far so good. The plants have survived and thrived. After they are harvested and studied, maybe the next group of astronauts will get to try out the fresh food. Maybe it will be you. You will have your very own fresh food market right on your spacecraft.

FOLLOW
the MONEY

Money talks. Where there is money, there is exploration. In the United States, NASA gets about $3.8 billion a year from the national government for space exploration. NOAA gets around $23.7 million a year for deep-sea exploration. That means that NASA gets 150 times what NOAA does. Why the big difference? Most people believe that space exploration requires more money. It costs more to reach other planets. Maybe so. But ocean exploration is not cheap. It cost James Cameron eight million dollars to take a one-man submersible to the Mariana Trench. The question is, which exploration will benefit humans more? The answer should be—both!

AN ARTIST SHOWS NASA'S NEW HORIZONS SPACECRAFT ON ITS WAY TO SEE PLUTO FOR THE FIRST TIME.

Did you know that in the last 40 years we have made some extraordinary discoveries in the ocean? In 1977, scientists noticed vents of hot water deep in the ocean floor. They were puzzled. The water in the ocean is normally close to freezing near the bottom. But there were areas of very warm water shooting up from holes, or vents, in Earth's crust. A whole new group of organisms never before seen were found to be living in and around these vents. The warm water is released from deep within the planet through these holes in the crust. The holes form as the tectonic plates shift and the pressure builds up, forcing liquid through the vents. Temperatures near the vents can reach 700°F (340°C). Yet the water does not boil. The extreme pressures at the vents make the water come out as bubbles or shifting currents.

The discovery of these vents changed how scientists viewed Earth. This confirmed their idea that Earth's crust comprised plates that move. It supported their ideas for how and why earthquakes, tsunamis, and volcanoes occur. These discoveries are all thanks to one deep-sea expedition near the Galápagos Islands. Makes you wonder how many other amazing discoveries lie beneath the deep.

EXPLORER'S NOTEBOOK

4 WHY AND HOW DO WE EXPLORE?

✓ Humans are curious. We want to know everything we can about our planet and the universe.

✓ Exploring the deep sea and deep space are ways to do that

✓ Don't forget your suit!

✓ Use probes, satellites, and AUVs to help explore.

✓ Learning about the ocean is as important as learning about space.

ACTIVITY:

DESIGN YOUR OWN SPACE SUIT

What would you need on your space suit? Draw a picture of the tools and safety equipment that you might need.

Draw a picture of a diving suit. Include the tools and safety equipment you might need.

COMPARE THE TWO. What's different? What is the same?

SUPPLIES

Drawing tools

Paper

Imagination

HERE ARE SOME IDEAS TO GET YOU STARTED

THINGS TO THINK ABOUT:

How long will you be out in space/ underwater?

What do you need to be able to do? Are you just walking? Do you need to move objects? What kinds of tools will you need?

How much air will you need?

What kind of forces will be acting on you?

CHAPTER 5

WHAT HAVE WE LEARNED?

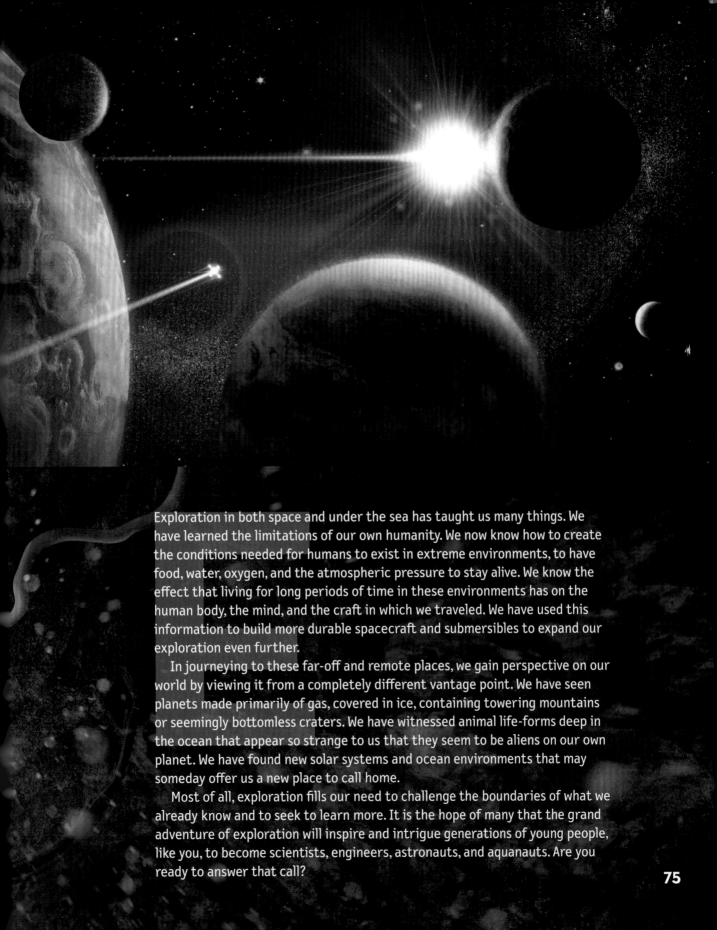

Exploration in both space and under the sea has taught us many things. We have learned the limitations of our own humanity. We now know how to create the conditions needed for humans to exist in extreme environments, to have food, water, oxygen, and the atmospheric pressure to stay alive. We know the effect that living for long periods of time in these environments has on the human body, the mind, and the craft in which we traveled. We have used this information to build more durable spacecraft and submersibles to expand our exploration even further.

In journeying to these far-off and remote places, we gain perspective on our world by viewing it from a completely different vantage point. We have seen planets made primarily of gas, covered in ice, containing towering mountains or seemingly bottomless craters. We have witnessed animal life-forms deep in the ocean that appear so strange to us that they seem to be aliens on our own planet. We have found new solar systems and ocean environments that may someday offer us a new place to call home.

Most of all, exploration fills our need to challenge the boundaries of what we already know and to seek to learn more. It is the hope of many that the grand adventure of exploration will inspire and intrigue generations of young people, like you, to become scientists, engineers, astronauts, and aquanauts. Are you ready to answer that call?

ASTRONAUT
OR AQUANAUT?

It's time to decide. Will you become an astronaut or an aquanaut? Can't make up your mind? Don't worry. It is possible to be interested in both fields. In fact, more and more researchers from space find themselves working with oceanographers and marine scientists. Information learned in either place can be helpful to the other. For example, scientists have spent many years researching Hawaii's volcanoes. They have learned about plate tectonics, how parts of the planet's crust can be made of pieces that shift and move. They have discovered how hot, liquid lava, called magma, flows and pushes against the plates. They have learned how volcanoes form and erupt. All of this information is extremely useful to planetary geologists, scientists who study the makeup of celestial bodies.

LAVA FLOWING INTO THE PACIFIC
OCEAN ON THE ISLAND OF HAWAII

Since they can't actually go to these far-off places, scientists use the concepts from the Hawaiian islands, and other volcanoes on Earth, to understand what might be happening on other planets or moons. This is especially true of Enceladus, one of Saturn's moons. Researchers have discovered evidence of what appears to be an icy volcano on its surface. It is called a cryo-volcano. How did they know what it was? They compared how it looked with the volcanoes on Earth. Studying island volcanoes in the ocean or even volcanoes on the ocean floor is a great way for researchers to learn about space.

PHOTOS CAPTURED BY NASA'S DAWN MISSION HELPED CREATE THIS SIMULATED IMAGE OF AHUNA MONS, A MOUNTAIN ON THE DWARF PLANET CERES.

THIS SPOTTED OCEANIC TRIGGERFISH HIDES IN THE MIDDLE OF PLASTIC GARBAGE.

GARBAGE Is EVERYWHERE

Not only do we have a garbage problem in our oceans—we also have one in space. In the ocean the garbage consists of bits of material—debris from shore or ships. It can be old shoes, discarded fishing poles, metal soda cans, the plastic rings that hold them together, or pretty much anything that people throw away. Regardless of the type of garbage, it's all dangerous to marine life. There is a similar problem in space, but this "space junk" is made of bits of old spacecrafts, satellites, and even tools that escaped from an astronaut's grasp. Even though there aren't animals in space, the garbage is still dangerous. Unless you think a bit of metal coming at you at speeds up to 17,550 miles an hour (28,224 km/h) isn't something to worry about. Garbage, whether in space or in our oceans, is something to be concerned about. It's time to clean it up! Add that to your list of things to do as an astronaut or aquanaut.

WHAT CAN THE OCEAN
TEACH US ABOUT SPACE?

Space exploration is about understanding how the universe works. It's also a quest to find life on other planets or moons. The ocean has taught us many things about how and what to search for in space. It has made us challenge beliefs that we have held about what is needed for life to exist.

First of all, life does not always require the sun. Life has been found in the darkest, deepest depths of the ocean. Sunlight does not reach down there. Living things can survive in cold temperatures, high pressures, and extreme toxicity. The bottom of the ocean is a harsh, frigid environment. It is filled with high metal levels and hydrogen sulfide, a toxic gas dangerous to humans. Living things can survive with little food. Creatures that live in the dark depths of the ocean do not have a large food source. They mostly eat each other or the tiny bits of food that sink to the bottom. There is very little oxygen at the bottom of the ocean, and yet life exists there.

Will humans be able to live in this environment? No. Thus, life on other planets will likely resemble these sea creatures more than humans.

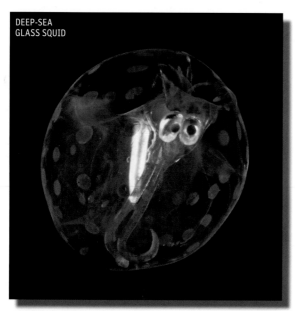

DEEP-SEA GLASS SQUID

ONE-FIN FLASHLIGHT FISH

DEEP-SEA ANGLERFISH

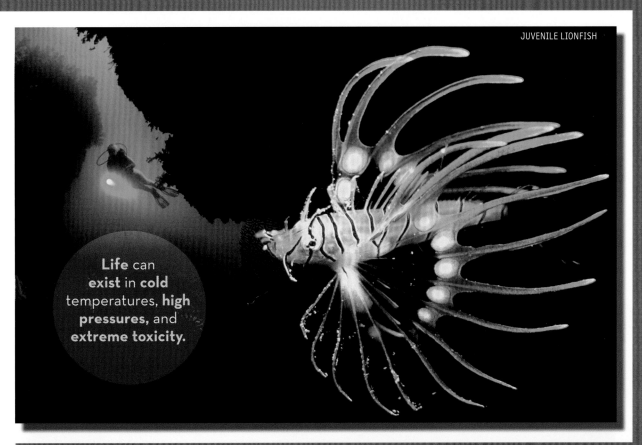

JUVENILE LIONFISH

Life can **exist** in **cold** temperatures, **high pressures,** and **extreme toxicity.**

How can STUDYING the OCEAN help ASTRONAUTS better UNDERSTAND CONDITIONS in SPACE?

When you are underwater, you become very aware of the freedom of movement. Water teaches you quickly to slow yourself down and make tiny adjustments instead of large ones. This type of movement is also needed in space. Small, focused movements can be used to create very large forces. By training underwater, astronauts will understand this and will store this information in their muscle memory. Astronauts call this reducing your gains. Smaller gains give you big results.
—Dr. Kathryn Sullivan, astronaut-aquanaut and former director of NOAA

ASTRONAUTS HEADING TO AQUARIUS FOR AN UNDERWATER TRAINING SESSION WITH PROJECT NEEMO

WHAT CAN SPACE TEACH US
ABOUT THE OCEAN?

Space and the deep sea are similar environments. They are hostile and dangerous to humans. The equipment that we have developed to survive in space is the same type we will need to survive deep underwater. Space suits, remotely operated vehicles, and spacecraft have designs that are useful to deep-sea researchers. Satellites are extremely helpful as well. They can provide information about the sea that is impossible to get from being on the land or even in the deep sea. Satellites offer a bird's-eye view of the ocean. They can be used to map the ocean floor, to show different temperatures, and to identify terrain such as deep trenches and high mountains. Satellites track carbon absorption and carbon dioxide emissions, which tell us how our oceans affect climate. They track the amount of algae that is present, an indication of the health of the

SATELLITES OFFER A BIRD'S-EYE VIEW OF THE OCEAN.

oceans. They show color changes in the ocean, which reveal how much plankton—the food required for whales, shrimp, snails, and jellies—is in the water. Plankton is an extremely important indicator for ocean and overall planet health since it is at the bottom of the food chain.

PLANKTON

A SATELLITE CAPTURED THIS INFRARED IMAGE OF HURRICANE PATRICIA, ONE OF THE STRONGEST HURRICANES TO HIT LAND.

DISCOVERIES THAT HAVE
CHANGED OUR LIVES

Maybe you want to make new discoveries—ones that could change how we live. Choosing to study space or the deep sea would be a great way to do that. Much of the technology developed for both types of exploration has become part of our daily lives. We owe the space program for helping to develop solar energy, an eco-friendly way to generate electricity from sunlight. Long-distance communication through satellites, which includes satellite television, came from the technology of space exploration. NASA contributed to the development of artificial limbs by creating robots for its spaceships. NASA helped develop smoke detectors, shoe insoles, infrared

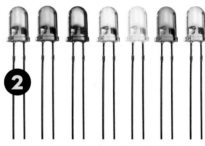

1 BREATHING EQUIPMENT USED BY FIREFIGHTERS WAS FIRST DEVELOPED BY NASA FOR USE IN SPACE.

2 LIGHT-EMITTING DIODES (LEDs) LIKE THESE WERE CREATED BY NASA TO HELP PLANTS GROW IN SPACE. ON EARTH, THEY ARE USED TO HELP HEAL SORE MUSCLES AND JOINTS.

3 CHEMICAL SENSORS BUILT TO HELP ASTRONAUTS DETECT WEAKENING METAL STRUCTURES IN SPACE ARE BEING USED AS ALARMS THAT DETECT TOXIC CHEMICALS ON THE EARTH.

4 IN PLANNING FOR LONG-TERM SPACE TRAVEL, NASA DEVELOPED A WAY TO PRESERVE FOOD BY FREEZE-DRYING IT. THE SAME TECHNOLOGY OF FREEZE-DRYING FOOD IS USED TO HELP PRESERVE FOOD ON EARTH (AND TO CREATE FUN SNACKS).

5 THIS SPIRAL BRIDGE, BUILT IN SHANGHAI, CHINA, IS A WAY OF MAKING THE BEST USE OF SPACE IN A CROWDED COUNTRY. SIMILARLY, ENGINEERS DESIGNING THE ISS AND OTHER SPACECRAFT MUST MAKE THE BEST USE OF THE VERY LIMITED SPACE AVAILABLE.

6 NASA USED THEIR WIND TUNNELS AND AERODYNAMIC RESEARCH TO HELP CREATE THESE NANOTECH SWIMSUITS THAT SPEED SWIMMERS THROUGH THE WATER IN RECORD-BREAKING TIMES.

ear thermometers, nanotechnology swimsuits, and even invisible braces. They have also helped create digital imaging in the form of magnetic resonance imaging (MRI) and computerized tomography (CT) scans for hospitals.

Space exploration is not the only frontier to greatly contribute to the field of medicine. Through ocean research on algae, marine animals, plants, and ecosystems, new drugs are being created. Since oceans contain more than 80 percent of the different plants and animals in the world, they may be just the place to find ways to fight human diseases. The ocean is a huge resource for scientists who are concerned about the effect climate change is having on our planet. They are studying the coral reefs, the aquatic animals and their habitats, and even water temperatures to find ways to combat climate change. After all, the ocean, like the forest and the desert, is one of the major ecosystems of the planet, and one that affects the billions of living organisms living here.

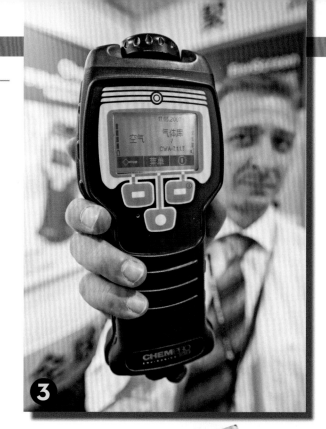
❸

❹

❺

❻

WORKING
TOGETHER

There is a great way to be an astronaut or aqua-naut and share ideas! Work on a joint NASA-NOAA project. NASA is the United States space agency and NOAA is the United States agency that handles deep-sea exploration. They both conduct research in very different places, but they find that they work best when they pool their information. The following are a few of the programs that they are working on together.

Interested in weather and climate control? Maybe one day you could work on the Joint Polar Satellite System (JPSS). JPSS is a new program that will help NOAA collect information on weather patterns, climate change, and overall global environment. It is made up of spacecraft, satellites, and a ground system all run jointly by NOAA and NASA. The goal is to improve advance weather notifications and make more accurate weather forecasts to keep people and their homes and businesses safe.

Want to learn more about El Niño? NASA and NOAA are teaming up on an El Niño storm mission. They will be using the Global Hawk, an unmanned plane, to collect data about the El Niño weather system. The remotely piloted plane is able to fly at 65,000 feet (20 km) for up to 30 hours at a time. A human-piloted plane would be unable to do this. The Global Hawk will be gathering data from its many sensors and passing that information to land bases. Researchers there will compile the data and use it to form predictions and forecasts for future weather occurrences. The flexibility of being able to stay in the air for longer periods of time and at different altitudes allows for much more detailed information than from a satellite.

Want to know about solar activity and how it affects weather? Check out the NOAA Deep Space Climate Observatory (DSCOVR) satellite. Launched in 2015, this newly renovated satellite contains space-weathering instruments that will track and record the activity of the sun. It will also be able to measure ozone and aerosols in the atmosphere as well as measure electrons in the solar wind. NASA was in charge of launching the satellite, but after 150 days, it turned over its operations to NOAA. If you work on this project, you may just become the first deep-space weather forecaster.

THE DEEP SPACE CLIMATE OBSERVATORY SPACECRAFT WILL MONITOR SOLAR ACTIVITY AND EARTH'S ATMOSPHERE.

ONE AGENCY
to **RULE** Them **ALL**

You have seen that there are many similarities between space and the deep sea. Both environments are mostly without light, hostile to humans, isolating, and yet filled with wonder. With NASA and NOAA doing research in two different areas, the question of combining them has come up more than once. Some people have suggested that the best use of research money is to create a joint NASA and NOAA agency. The money could be used to further research in both areas. After all, it is clear that information gathered from both places has already been helpful. With more money, ocean exploration could be expanded. We may be able to discover new species and ecosystems right on our own planet. Perhaps those discoveries will help us understand the environments on Europa, Mars, Jupiter, and other planets and moons.

By pooling our information, we may be able to come up with ways to help keep our planet healthy, and to use our knowledge of what happens on other planets to prevent that from happening here. NASA and NOAA researchers would have access to all of the information gathered so that joint research could be conducted. Who knows where this will lead? Maybe we will discover other planets or moons like ours, and also aid in the search for life in the universe. The possibilities are endless.

Maybe we will **discover** other **planets or moons** like ours, and also aid in the **search for life** in the universe.

TO TEST SPACE EQUIPMENT ON EARTH, NASA CREATED THIS THERMAL VACUUM CHAMBER TO MIMIC THE VACUUM OF SPACE.

SPACE COLONIES AND
UNDERWATER CITIES

As the population of the Earth keeps growing, there has been a lot of discussion about where to expand. Should we build space stations on the moon? Should we have a colony on Mars? Perhaps it would be better to build underwater communities or cities? The only way to decide is to research and invent new technology that would make this possible. Living under the sea would keep people safe from violent weather systems, tsunamis, and even severe temperature swings. It would allow humans a unique view of underwater life. Living in space would be a new adventure, one that might involve new temperatures, weather systems, and environments. It would expose humans to a new view of space and the sky and would offer them the chance to explore a new planet or moon.

The prospects are endless. The idea may seem far-fetched, but it is one that's being discussed—even planned for. Which new world sounds appealing to you?

CONCEPT CITY ON MARS

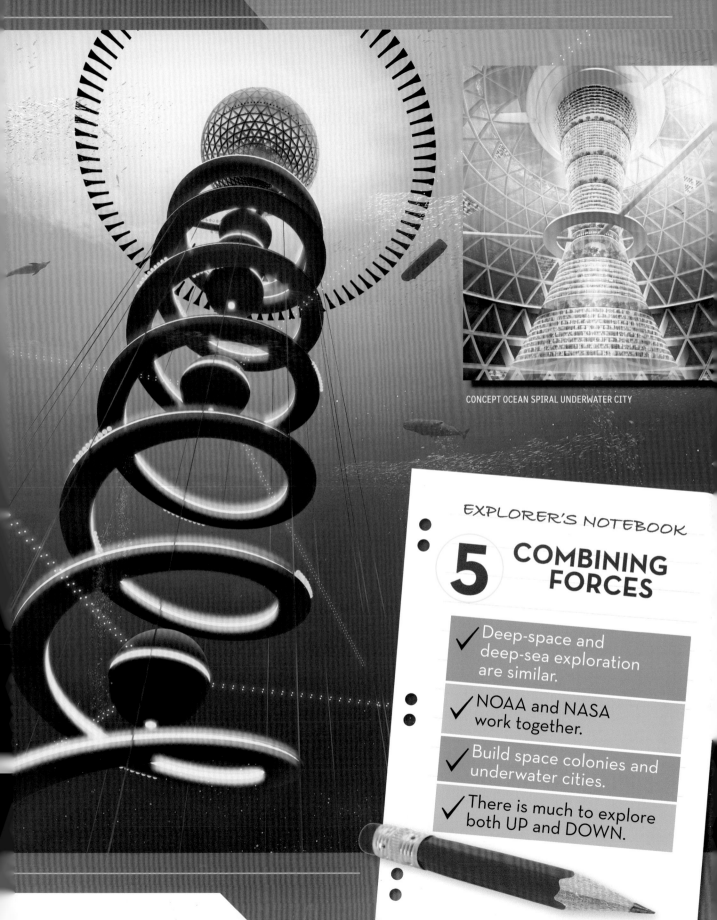

CONCEPT OCEAN SPIRAL UNDERWATER CITY

5 COMBINING FORCES

✓ Deep-space and deep-sea exploration are similar.

✓ NOAA and NASA work together.

✓ Build space colonies and underwater cities.

✓ There is much to explore both UP and DOWN.

MEET THE
ASTRONAUTS AND AQUANAUTS

DR. CATHERINE "CADY" COLEMAN

Dr. Catherine "Cady" Coleman is a former United States Air Force officer, and a current NASA astronaut. She is a veteran of two space shuttle missions, and she departed the International Space Station on May 23, 2011, as a crew member of Expedition 27 after logging 159 days in space. Dr. Coleman holds a Ph.D. in polymer science and engineering.

DR. ANDREW ABERCROMBY

Dr. Andrew Abercromby has fourteen years' experience as an engineer, scientist, and research diver at NASA's Johnson Space Center in Houston, Texas, working on the design and testing of spacecraft and space suits. He is the lead of NASA's EVA Physiology Laboratory (EVA = space suits) and also serves as EVA Scientist for the Biomedical Research and Environmental Sciences Division, which means that he tries very hard to coordinate space-suit-related research activities across a lot of different laboratories and tests.

FABIEN COUSTEAU

As the first grandson of Jacques-Yves Cousteau, Fabien Cousteau spent his early years aboard his famous grandfather's ships, learning how to scuba dive when he was very young. Today, he works as an aquanaut and a documentary filmmaker to protect and preserve the planet's extensive and endangered marine inhabitants and habitats.

DR. BRIAN HELMUTH

Dr. Brian Helmuth is a professor of marine and environmental science and public policy at Northeastern's Marine Science Center. His research explores the effects of climate and climate change on the physiology and ecology of marine organisms. He has participated in several projects on Aquarius and has served as science director for many underwater missions.

DR. DARLENE LIM

Dr. Darlene Lim is a geobiologist based at the NASA Ames Research Center. She is the principal investigator for the BASALT and PLRP programs, tasked with identifying best practices and developing operational concepts for the human scientific exploration of our solar system. Darlene has spent more than 20 years leading field research programs around the world. She has conducted research in both the Arctic and Antarctic, and in various underwater environments where she has spent hours piloting submersibles as a scientist and explorer.

DR. KATHRYN SULLIVAN

Dr. Kathryn Sullivan is a geologist and a former NASA astronaut. A crew member on three space shuttle missions, she's the first American woman to have walked in space. She's the former under secretary of commerce for oceans and atmosphere and NOAA administrator.

DR. NORMAN THAGARD

Dr. Norman Thagard is a retired NASA astronaut and medical doctor. A veteran of five space flights, he has logged more than 140 days in space. He was a mission specialist on STS-7 in 1983, STS 51-B in 1985, and STS-30 in 1989, and was the payload commander on STS-42 in 1992. He was the cosmonaut/researcher on the Russian Mir 18 mission in 1995.

ELISABETH MAGEE

Elisabeth Magee is program coordinator of the Three Seas Program, in addition to serving as the diving safety officer for Northeastern University and as an aquanaut with the Mission 31 program. She has been a research diver for 10 years, beginning as an undergraduate student in the Three Seas Program at Northeastern University, and now has logged more than 1,000 research dives. Elisabeth is an active AAUS Scientific Diving and NAUI Scuba instructor. She enjoys teaching scientific diving to the next generation of researchers at Northeastern University's Marine Science Station and is passionate about helping young divers enter the field as blossoming professionals.

DR. DAWN WRIGHT

Dr. Dawn Wright is professor of geography and oceanography at Oregon State University and the first woman of color to dive in the three-person autonomous craft Alvin in 1991. An ocean scientist, geographer, leading authority in geographic information system (GIS), and author, Dawn is the chief scientist at Esri. She continues to enjoy studying the ocean through the windows of a deep-ocean submersible.

DR. MICHAEL VECCHIONE

Dr. Michael Vecchione is the director of the NOAA Fisheries National Systematics Laboratory and is assigned to work with the National Museum of Natural History. His research is on cephalopod (squids and octopuses) natural history and biodiversity. He spends a lot of time at sea either leading or participating in deep-sea and polar expeditions on U.S. and foreign ships using both submersibles and traditional net sampling to study deep-sea animals.

SPACE AND DEEP SEA
MORE ALIKE THAN YOU IMAGINE

THE AIRCRAFT CARRIER
U.S.S. *ENTERPRISE*

F/A-18E SUPER
HORNET

STAR WARS
TIE FIGHTER

SPACESHIPS VS.
SHIPS on the WATER

Whether it's a ship on the water or a ship in space, they tend to have the same names. Why? They are often taken from ships from the U.S. Navy. After all, ships on the water are known as boats, cruisers, battleships, or craft. Add a "space" in front of these and you'll find a lot of the names for spaceships. It even works for planes: Star Wars TIE fighter vs. U.S. Navy F/A-18 fighter jet. But not every boat name works. Have you ever seen a space kayak?

BRIDGE on a SPACESHIP
VS. BRIDGE on a SHIP

A SHIP'S BRIDGE

The bridge of a ship is the seat of command. It is where the captain stands and gives orders. It is where all of the people who are in charge of steering the ship and defending it can be found. A bridge on a spaceship is an open space with areas for electronic consoles and a huge window to look outside, or out in space as the case may be.

SPACE SHUTTLE *ATLANTIS* FLIGHT DECK

ALIENS IN SPACE
VS. ALIEN DEEP

Both space and deep under the ocean are foreign territory to humans. Neither is a place that we go very often or easily. Because of this, we think of the inhabitants of these places as "aliens."

AMERICAN PADDLEFISH

FURTHER STUDY
WEBSITES
FOR KIDS

National Geographic
nationalgeographic.org/activity/
exploring-extremes

NASA website showing the space suits throughout the years:
nasa.gov/externalflash/nasa_spacesuit

NOAA Ocean Explorer for Kids
oceanservice.noaa.gov/kids

GLOSSARY

Asteroid—a small rocky object, smaller than a planet, that orbits the sun

Atmosphere—the layers of gas that surround a planet

Ballast tank—a tank located in a floating structure that can take on water to add weight to the object and make it sink, or release water to make it float

Bioluminescence—the ability of an organism to create light from its own body

Black hole—a spot in space where the gravity is so intense, objects are pulled into it and cannot escape

Buoyancy—the ability to float

FRESH IMPACT CRATER ON MARS

Crater—a large, round, or oblong indented area on the surface of a planet or moon

Density—the mass per unit volume of a substance

Earthquake—violent shaking of the ground that occurs suddenly, usually due to shifting of tectonic plates

Free fall—the falling of an object because of gravity

Gravity—the force of attraction between two objects such as planets, stars, or moons. It is also the force felt on Earth that pulls everything toward the center.

Magma—extremely hot or molten rock

Meteor—a rock or bit of a star or planet that enters Earth's atmosphere

Microgravity—the condition where objects appear to be weightless or floating

Navigation—the process of determining where you are and steering toward where you want to go

Orbit—a regular, curved path around a celestial body

Orbiter—a spacecraft designed to go into space

Ozone layer—the layer in the stratosphere that contains ozone, a gas that absorbs ultraviolet light and keeps the Earth cool or hot

Plankton—microscopic organisms that float in freshwater or the ocean

Precipitation—a form of rain, snow, sleet, or hail that drops from clouds to the ground

Pressure—a continuous force exerted on an object

Satellite—an object that orbits around a celestial body; can be natural or man-made

Submersible—a boat or craft that is meant to operate under the water

Tectonic plates—enormous slabs of rock that make up the Earth's crust

Topography—the physical features of the surface of a celestial body

Tsunami—a long sea wave that is formed due to an underwater volcano, earthquake, or other disturbance

Vacuum—in space, the almost complete absence of pressure

Weightlessness—the state of floating or not having gravity act upon an object

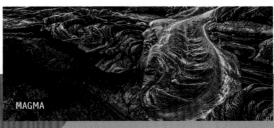

MAGMA

CREDITS

Cover: Antonio Caparo; dust jacket flap: (Jennifer Swanson), Jeff Olson; (Fabien Cousteau), Carrie Vonderhaar; (Kathryn Sullivan) NOAA

Front matter: 1, Tammy616/E+/Getty Images; 2-3 (UP), Alan Uster/Shutterstock; 2-3 (LO), Ethan Daniels/Shutterstock; 4 (UP), Andrey Armyagov/Shutterstock; 4 (LO), Tim Laman/National Geographic Creative; 6-7 (UP), Jesse Kraft/Alamy Stock Photo; 6-7 (LO), Richinpit/E+/Getty Images; 8-9 (UP), NASA; 8-9 (LO), Kyle McBurnie

Chapter 1: 10-11, Antonio Caparo; 12, takau99/Moment RF/Getty Images; 13 (UP), NASA; 13 (LO LE), ESA/Hubble & NASA; 13 (LO RT), E. Widder/HBOI/Visuals Unlimited/Getty Images; 14, NASA; 15 (UP), NASA; 15 (LO), Natalin*ka/Shutterstock; 16, IPGGutenbergUKLtd/ iStockphoto/Getty Images; 17, Lachina; 18 (LE), Philippe Poulet/ Mission/Taxi/Getty Images; 18 (RT), NASA; 19, NOAA; 20, Alex Mustard/Nature Picture Library; 20 (INSET), Dana Stephenson/ Getty Images; 21 (UP), Mark Garlick; 21 (LO), David Aguilar; 21 (LO RT), Hans F. Meier/iStockphoto/Getty Images; 22-23, JPL/ University of Arizona/NASA; 23 (red pencil), rvlsoft/Shutterstock; 24 (paper lightbulb), Brian A Jackson/Shutterstock; 24-25 (pen cap), Sergej Razvodovskij/Shutterstock; 24-25 (water bottle), Zonda/Shutterstock; 24-25 (clay), oksana2010/Shutterstock; 24-25 (bottle cap), Zonda/Shutterstock

Chapter 2: 26-27, Antonio Caparo; 28, JSC Image Repository/Terry Slezak/NASA; 29, V. Crobu/ESA; 30, Karl Shreeves/NASA; 31, Stuart Armstrong; 32 (LE), Tim Laman/National Geographic Creative; 32 (CTR), Tim Laman/National Geographic Creative; 32 (RT), Tim Laman/National Geographic Creative; 33, Tim Laman/National Geographic Creative; 35 (UP LE), NASA; 35 (UP RT), NASA; 35 (LO LE), solarseven/iStockphoto/Getty Images; 35 (LO CTR), IM_photo/ Shutterstock; 35 (LO RT), John Pitcher/iStockPhoto; 36, Original artwork by Francis J. Krasyk, adapted by Stuart Armstrong; 37, Kelvin Aitken/VWPics/Alamy Stock Photo; 38-39, NASA; 39, Bill Ingalls/NASA; 40 (sneakers), arka38/Shutterstock; 40 (pants), Karkas/Shutterstock; 40 (underwear), Elnur/Shutterstock; 40 (socks), alexandre zveiger/Shutterstock; 40 (teddy bear), Ovydyborets/Dreamstime; 40 (pocket knife), Ingram; 40 (iPod), You can more/Shutterstock; 40 (wet suit), Photodisc; 40 (LO), SPUTNIK/Alamy Stock Photo; 41 (red pencil), rvlsoft/Shutterstock; 42 (spool of rope), Hayati Kayhan/Shutterstock; 42-43 (red cup), Joe Belanger/Shutterstock; 42-43 (tennis ball), Zheltyshev/ Shutterstock; 42-43 (eye screw), FCG/Shutterstock; 43 (paper light bulb), Brian A Jackson/Shutterstock; 43 (knotted rope), Elnur/Shutterstock; 43 (child's waist), Jacek Chabraszewski/ Shutterstock

Chapter 3: 44-45, Antonio Caparo; 46, NASA; 47, NASA; 48-49, Kip Evans/Alamy Stock Photo; 49 (LE), NASA; 49 (RT), NASA; 50, NASA; 51 (LE), Robyn Beck/AFP/Getty Images; 51 (RT), NASA; 53 (UP), D.Ducros/ESA; 53 (LO), NASA; 54, NASA; 55, NASA; 56, Bill Ingalls/ NASA; 57 (red pencil), rvlsoft/Shutterstock

Chapter 4: 58-59, Antonio Caparo; 60, E. Bacon/Topical Press Agency/Getty Images; 61 (UP), Fox Photos/Getty Images; 61 (LO), The U.S. National Archives and Records Administration; 62, Bill Stafford/NASA; 63 (UP LE), AFP/Getty Images; 63 (UP RT), Topical Press Agency/Getty Images; 63 (LO), NASA; 64, NASA; 65, (BOTH) Nuytco Research Ltd.; 66-67 (LO), NOAA; 67 (UP), NASA; 68, NASA; 69, Emory Kristof/National Geographic Creative; 70, NASA; 71 (UP), NASA; 71 (red pencil), rvlsoft/Shutterstock; 72 (colored pencils), Artem Shadrin/Shutterstock; 73, Kaya Dengle

Chapter 5: 74-75, Antonio Caparo; 76, Allen.G/Shutterstock; 77 (UP), JPL-Caltech/UCLA/MPS/DLR/IDA/PSI/NASA; 77 (LO), Paulo de Oliveira/Newscom; 78 (UP), David Shale/Nature Picture Library; 78 (LO LE), Norbert Wu/Minden Pictures; 78 (LO RT), Peter David/ The Image Bank/Getty Images; 79 (UP), Reinhard Dirscherl/Alamy Stock Photo; 79 (LO), NASA; 80-81, Romolo Tavani/Shutterstock; 81 (UP), Al Giddings Images; 81 (LO), William Straka III, University of Wisconsin-Madison, CIMSS/NASA; 82 (LE), Flashon Studio/ Shutterstock; 82 (RT), Krasowit/Shutterstock; 83 (UP), Teh Eng Koon/AFP/Getty Images; 83 (LO LE), ArtisticPhoto/Shutterstock; 83 (CTR), Hugh Threlfall/Photolibrary RM/Getty Images; 83 (LO RT), PA Images/Alamy Stock Photo; 84, NASA; 85, NASA; 86, David Aguilar; 87 (BOTH), Ocean Future City Concept Project/Shimizu Corporation; 87 (red pencil), rvlsoft/Shutterstock

End matter: 88-89 (UP), NASA; 88-89 (LO), David Doubilet/ National Geographic Creative; 88 (Dr. Cady Coleman), NASA; 88 (Dr. Brian Helmuth), Courtesy Dr. Brian Helmuth; 88 (Fabien Cousteau), Carrie Vonderhaar; 88 (Dr. Andrew Abercromby), Courtesy Dr. Andrew Abercromby; 89 (Dr. Kathryn Sullivan), NOAA; 89 (Dr. Norman Thagard), NASA; 89 (Dr. Dawn Wright), Courtesy Dr. Dawn Wright; 89 (Elisabeth Magee), Courtesy Elisabeth Magee; 89 (Dr. Darlene Lim), Scott S. Hughes; 89 (Dr. Michael Vecchione), Courtesy Dr. Michael Vecchione; 90 (LO), Mark Conlin/Alamy Stock Photo; 90 (UP RT), CTR Photos/ iStock/Getty Images; 90 (CTR LE), deaw59/iStockphoto/Getty Images; 90 (UP LE), U.S. Navy photo by Petty Officer 1st Class Todd Cichonowicz/U.S. Navy; 90 (UP LE INSET), U.S. Navy, Official Photograph; 90 (CTR RT), NASA; 91 (UP), NASA; 91 (LO), Willyam Bradberry/Shutterstock; 96, Julie Swanson

INDEX

Illustrations are indicated by **boldface.**

INDEX

Since 1888, the National Geographic Society has funded more
than 12,000 research, exploration, and preservation projects
around the world. The Society receives funds from National
Geographic Partners, LLC, funded in part by your purchase.
A portion of the proceeds from this book supports this vital
work. To learn more, visit natgeo.com/info.

NATIONAL GEOGRAPHIC and Yellow Border Design are trade-
marks of the National Geographic Society, used under license.

For more information, visit nationalgeographic.com,
call 1-800-647-5463, or write to the following address:
    National Geographic Partners
    1145 17th Street N.W.
    Washington, D.C. 20036-4688 U.S.A.

Visit us online at nationalgeographic.com/books

For librarians and teachers: ngchildrensbooks.org

More for kids from National Geographic:
kids.nationalgeographic.com

For information about special discounts for bulk purchases,
please contact National Geographic Books Special Sales:
specialsales@natgeo.com

For rights or permissions inquiries, please contact National
Geographic Books Subsidiary Rights: bookrights@natgeo.com

Library of Congress Cataloging-in-Publication Data

Names: Swanson, Jennifer. Title: Astronaut, aquanaut/by
    Jennifer Swanson.
Other titles: Aquanaut
Description: Washington, DC : National Geographic Kids, 2018. |
    Audience: Age 9-12. | Audience: Grade 4 to 6. | Includes index.
Identifiers: LCCN 2017020436| ISBN 9781426328671 (hardcover)
    | ISBN 9781426328688 (hardcover)
Subjects: LCSH: Outer space--Exploration--Juvenile literature.
    | Underwater exploration--Juvenile literature.
Classification: LCC TL793 .S9425 2018 | DDC 627/.72--dc23
LC record available at https://lccn.loc.gov/2017020436

Printed in China
17/RRDS/1

AUTHOR JENNIFER SWANSON WITH AQUANAUTS FABIEN COUSTEAU AND ELISABETH MAGEE. FABIEN SHOWS OFF THE *ASTRONAUT-AQUANAUT* BOOK COVER ON HIS PHONE.

To astronaut Sally Ride and aquanaut Jacques Cousteau, the two heroes of my youth. You both helped me discover the amazing wonder of science in the worlds above and below. —JS

Acknowledgments: Many thanks to all the astronauts and aquanauts who shared your experiences with my readers.
Fabien Cousteau
Dr. Kathryn Sullivan
Dr. Brian Helmuth
Dr. Darlene Lim
Dr. Dawn Wright
Elisabeth Magee
Dr. Norman Thagard
Dr. Michael Vecchione
Dr. Cady Coleman
Dr. Andrew Abercromby

This book is so much richer for your contributions! Special thanks to Dr. Brian Helmuth and Dr. Darlene Lim for your expert review of the book. Finally, big kudos to my amazing editor, Shelby Alinsky, who provided the spark of an idea that led to this fabulous book.

The author and publisher also wish to thank the book team: Shelby Alinsky, Kathryn Williams, Julide Dengel, Lori Epstein, Antonio Caparo, and Dawn Ripple McFadin.